THE TECHNICAL WRITING CASEBOOK

Thomas N. Trzyna
Seattle Pacific University

Margaret W. Batschelet
University of Texas at San Antonio

Wadsworth Publishing Company
Belmont, California
A Division of Wadsworth, Inc.

English Editor: John Strohmeier
Editorial Assistant: Holly Allen
Production Editor: Robin Lockwood, Bookman Productions
Designer: Hal Lockwood
Copy Editor: Ruth Cottrell
Compositor: Bookends Typesetting
Cover: Hal Lockwood

Printed in the United States of America

1 2 3 4 5 6 7 8 9 10 —— 92 91 90 89 88

ISBN 0-534-08658-6

Library of Congress Cataloging in Publication Data

Trzyna, Thomas N., 1946–
  The technical writing casebook.

  Includes index.
  1. Technical writing—Case studies.   I. Batschelet,
Margaret.   II. Title.
T11.T747 1988   808.0666021   87-23108
ISBN 0-534-08658-6

Preface

The Technical Writing Casebook presents thirty short, classroom-tested writing cases in a text that can be used as a supplement to *Writing for the Technical Professions* or another comprehensive technical or professional writing text. The opening chapter of *The Technical Writing Casebook* reviews the basics—the analysis of audience and purpose, problem solving, research methods, oral presentations, and visual aids. Chapter 2 explains how to write a report or letter in response to a case by following the entire writing process from assignment to final draft. Each of the following chapters focuses on a single type of document: business documents, types of reports, instructions and manuals, proposals, and brochures. In every chapter the cases are preceded by a short review of the featured formats.

The cases demand varying amounts of background research, problem solving, and audience analysis. They can be readily modified to meet a student's or instructor's need for assignments that are easier, more demanding, or focused on only one aspect of the writing process. No math is required, but the density and complexity of the data vary, as does the need for careful selection and summary of data for nontechnical audiences. Most assignments require students to design visual aids and to integrate these visual aids with the text of their reports. Many of the cases are appropriate for group assignments.

Two or more assignments accompany most cases, and the assignments often require students to adopt different roles and to write at different stages of developing situations.

Cases have been drawn from a variety of technical and social science fields, including engineering, urban planning, the health sciences, computer science, management, safety engineering, and agriculture. *The Technical Writing Casebook* has been tested in both technical writing workshops and traditional lecture courses. The cases have been well received by several levels of students, including traditional college sophomores, juniors, and seniors, as well as returning adult learners.

We wish to acknowledge the help of the following individuals, both friends and students, in developing material for these cases: Bennett Anderson, William Batschelet, Jerry Cook, George Fernandez, Todd Folz, Jeanne Geiger, Perry Grosch, Roxanne Howard, Michele Kennison, Adrien N. Martinez, Cynthia Myers, Vicky Rugh, Blake Stapper, and Jacqueline Tregre.

We would also like to thank the following reviewers: Dorothy Bankson, Louisiana State University; Don Richard Cox, University of Tennessee; Treva Edwards, Front Range Community College; Ruth E. Falor, Ohio State University; Claude L. Gibson, Texas A & M University; Robin McNallie, James Madison University; Marion K. Smith, Brigham Young University; and Judith Stanton, Clemson University.

Contents

Chapter *1*

STRATEGIES FOR TECHNICAL WRITING

INTRODUCTION

This chapter is designed to be a handbook of the basic strategies of technical writing: analyzing audience and purpose, problem solving, research, oral presentations, and visual aids. If you are new to technical writing, you will want to refer to this chapter often as you develop responses to the problems posed in the cases. Even if you have taken a technical writing course before, you will find this chapter a convenient reference. We begin by defining the characteristics of technical writing.

CHARACTERISTICS OF TECHNICAL WRITING

Technical writing has several characteristics that distinguish it from much of the personal and college writing you do. Technical writing is objective in tone; it is usually assigned like college writing; it responds to a particular need or purpose; it is usually designed to solve a specific problem; it often calls for highly technical vocabulary; and it is often done in cooperation with other writers. In addition, technical writing uses formats that are not common in other contexts, and it is more common in certain fields of work. Good technical writing, however, shares many characteristics with other good writing. The style and content of technical documents should be simple wherever simplicity is possible. Good technical writers also keep their purposes—and the needs of their audiences—in mind as they write. The next few sections of this chapter provide a brief overview of the major prewriting strategies used by professional technical writers—strategies for assessing audience and purpose, solving problems logically, and organizing well. As you read these sections remember that every technical writing task involves two parts—a *technical problem* that needs to be solved and a *communication problem* that needs to be carefully analyzed. Both tasks are equally important. If you cannot communicate your results, there is little point in solving technical or design problems.

AUDIENCE ANALYSIS

Your audience needs answers to specific questions, and it expects to read information that is written at a level it can understand. Audience analysis techniques are designed to help you to produce reports and letters that will satisfy your readers by answering their questions and by making them feel at home with the vocabulary and technical content of your presentations.

Audience Classification

The basic rule of thumb for audience analysis makes use of a simple classification of audiences. Your audience will usually include some combination of general readers, technicians, and experts.

The experts, in turn, can be divided into technical experts, managers, and executives. You can make finer discriminations about your audience by assessing its objectives, background knowledge, vocabulary, understanding of graphics and mathematics, and its expectations for organization and style. The following lists briefly summarize the items you need to consider for various audiences.

General Readers

Objectives: Assume the readers want general information, information that can be transferred to another field, and/or technical information in an easily understood form.

Background knowledge: Assume that background knowledge is limited.

Language: Use simple language with limited technical vocabulary.

Graphics: Use simple graphs (line, pie, bar) and photographs.

Measurements and relationships: Explain technical measurements by analogy or simplify them. Limit your use of statistics.

Organization and style: Use simple sentences with an average of 20 words per sentence. Use subtitles to develop and maintain interest.

Technicians

Objectives: Assume the readers want to find out how to do a task and/or learn how to explain a task to others.

Background knowledge: Assume they know their own fields and a limited amount about related fields. For example, an electrician needs to know about blueprints, property maps, and easements. Carefully consider how much a technician knows about related fields.

3

Language: Use simple language, but use technical terms where they are appropriate.

Graphics: Use complex graphics as long as they are types that are used in the specific field.

Organization and style: Use direct, imperative sentences. Use many informative subtitles for easy reference.

Technical Experts

Objectives: Assume the readers want information about a specific topic or information that can be applied to a related issue. For example, a wheelchair designer might read about aircraft hydraulics to look for transferable technologies.

Background knowledge: Assume they know their own fields well, but their specializations tend to be narrow. All engineers, for example, are not one audience. Electrical engineers know a different subject from mechanical engineers.

Language: Use plain language when there is no good reason to use technical vocabulary. Otherwise there is no limit on technical vocabulary or content.

Organization: Organize for efficient use. You should often use a standard academic format, with abstract, statement of problem, review of research, body, summary, conclusions and recommendations, and appendices for additional data and graphics.

Executives and Managers

Objectives: Assume the readers want information necessary for decision making. Executives tend to focus on long-term issues, managers on shorter-term issues. Both managers and executives need to make decisions based on cost/benefit analyses. Organize comparisons to highlight both financial costs and commitments of personnel.

Background knowledge: Assume they know their own fields well. Foreign material will need careful explanation.

Language and graphics: Use the same strategy as for technical experts. Keep in mind that these readers have little time to spend reading or searching for information.

Organization: Organize for efficient use. Include an executive summary and clearly labeled body, conclusions, recommendations, and appendices.

Questions to Ask about Your Audience

When you begin to write for a new audience, you can begin to identify the characteristics of that audience by asking the following questions:

1. What are my audience's expectations for language? Vocabulary? Technical terms?
2. How much can I assume that they know about mathematics?
3. What types of graphics are they used to seeing in their reports?
4. What measurement systems are familiar to them?
5. What are their primary fields of expertise? Have they also worked in other fields?
6. How much will they know about my field? About subjects related to their main field of expertise?
7. What are their objectives in reading my report?
8. How will they expect to see this report organized? What formats are they accustomed to seeing?

Once you recognize that many of your reports and presentations will be written for several audiences, you may need to work through these questions several times for a single report.

Multiple Audiences

Usually you will be writing for several audiences at the same time. At a minimum, you will write for your current audience and some future audience that will need to refer to your reports. One solution to the multiple audience problem is to develop the ability to adjust vocabulary, technical content, and graphics to meet the needs of several audiences simultaneously. Another solution is to organize a report with an introductory section that is suitable for all audiences and a series of subsections for each different audience. If you are writing to an audience composed of people you know personally, your task is simplified to the extent that you can use an Audience Profile Form.

Profiling an Audience

If you are writing a report to a person you know, you can obtain that person's profile by using the form in Figure 1-1. Instructions on the form itself explain how to answer the various questions.

When you have used one or more of these techniques to analyze your audience, you are ready to consider your purposes in writing a report or preparing an oral presentation.

DETERMINING PURPOSES

Your decisions about the purposes you want to achieve when you write or speak are closely related to audience analysis and equally as important. You will usually have a number of purposes. At a minimum, you will need to provide a record for later reference

Audience Profile Form

1. Name: (Enter the name of the individual.)

2. Role in the organization: (Enter the job title and other information about the person's function.)

3. Audience type: (general reader, technician, technical expert, manager, executive.)

4. Decision to be made: (What decision will this person make about your report?)

5. Audience level: (Do you consider this person a primary audience--a person who must make a key decision--or a secondary audience--a person who needs to be informed but is less involved in approving or rejecting your report?)

6. Idiosyncracies: (Does this reader prefer memos or executive summaries? Does he or she have a preference for certain kinds of technical content or vocabulary? For a certain management style?)

7. Receptiveness: (Does this reader normally read reports in the elevator? Late at night? What do you know about this reader's current successes or failures on the job that might affect your request or report?)

Figure 1-1. Audience profile form

in addition to your other purposes. Whenever you have several purposes, organize to meet your most important purpose first, your second most important purpose second, and so on. Seven basic purposes of reports, with a few notes on what you need to do to achieve each of these purposes, follow.

Directing Action

If your purpose is starting or ordering action, explain *what* is to be done, *when* it is to be done, and *how* it is to be done, and *by what organization or person* it is to be done.

Coordinating

Coordinating documents provide information for the successful management and completion of projects. Coordinating memoranda explain *who* does which parts of a project, *what* is being done, *when* meetings and deadlines occur, and *where* resources are located. Coordinating documents generally do *not* explain the reasons for the project.

Proposing and Requesting

Proposals explain *who* will do the work, *how much* money or assistance will be required and how that assistance will be managed, *when* the project will be undertaken, and *why* the proposal should be accepted. Proposals should also explain *how* the proposed action will benefit the agency that approves or funds the request.

Recommending

A recommendation is a suggestion, not an order or a request. Recommendations focus on alternatives. When you write a recommendation, *specify* the recommended action, *name* and discuss the alternatives, *present arguments* to support your choice, and *leave the final choice* to the reader who asked for the recommendation.

Providing a Record

A secondary purpose of most reports is to provide a record. If you need to provide a record, as in a progress report, *describe* important research in detail, covering who, what, when, where, and why; *discuss* any problems that occurred; *explain* decisions, solutions to problems, and changes in the original plans and procedures; and *indicate where records are filed.*

Informing

Purely informative reports have no expectations for action. If you want your audience to *do* something, consider whether you need to write a report, letter, or memo that coordinates, proposes, recommends, or serves some other purpose. Write an informative report when you want a co-worker to be aware of some development that might be of interest.

Entertaining

If your reports are entertaining, you will achieve two goals. The audience will pay attention, and the audience will tend to remember your material better. Clever graphics or humorous writing may *reinforce* important points and help to *simplify* complex information.

When you have analyzed your audience and decided on your purposes, you have the information necessary to make excellent decisions about organization. Yet before you can organize any data, you need to gather the data. Getting the data requires two steps: You need to define the problem set by your work assignment, and you need to perform the necessary research and solve your problem. The next two sections of this chapter address those steps in the writing process.

PROBLEM SOLVING

Analyzing your audience and identifying your purposes helps you to define the problem. At some point, however, you may also wish to *formulate a hypothesis or write out the questions you need to answer to solve the problem.* If you were asked to compile a personnel manual for a corporation, for example, you might list specific questions that you needed to answer before you could begin to organize and write. Those questions might include whether a personnel manual is a legal document, whether it constitutes part of an employment contract, and how other corporations strike a balance between producing a document for lawyers and a manual that any employee might understand.

Your next step is to *develop criteria for testing your hypothesis or answering your questions.* What kind of evidence can you use? What are the appropriate authorities in this case? How much evidence do you need before you can make a decision and get on with the rest of your job?

Next, of course, you will want to *apply your criteria* and reach a decision—assuming that you can find sufficient data. Many of the cases in this text provide examples of two common difficulties in problem solving. Sometimes you have trouble finding the significant data in a welter of extraneous material, and sometimes you must make a decision on the basis of insufficient data. If businesses always had all the data they wanted, there would never be any risk in new designs, new market plans, or new products. Your instructor may encourage you to do additional research to increase your probability of making the correct analysis, but you should be prepared to refine your criteria so that they can help you make decisions when you simply cannot find all the data you would like to have. The next section presents a few basic rules about research and information gathering.

RESEARCH METHODS

Professional researchers—market analysts, industrial information specialists, research librarians, and others—know that an effective search for information combines both field and library research. A good search also aims to find the most current information and includes interviews or correspondence with experts who can provide overviews, predictions for future trends, and criteria for decision making. Your searches for information should start with three basic questions: (1) Where is information about this topic likely to be found? (2) Who will probably know about this?, and (3) Who would pay for this information to exist?

Searching for Information

Thinking about the first question (Where is information about this topic likely to be found?) will probably lead you to several sources. Academic and special libraries are one source, and so are corporation libraries and the offices of professional societies, lobbies, and government agencies. If you are researching auto-emissions laws, you may find what you want by consulting indexes in a college library, but you may find more current information by calling an auto manufacturer's lobby or the office of a congressional committee.

The second question (Who will probably know about this?) leads to the same general answers as the first. When you think about information as something that resides in persons, not in books, you are stimulated to begin a search by asking a local expert for the names of other experts who may be able to answer your questions directly—and perhaps even make you aware of developments that will not be in print for months or even years.

The third question (Who would pay for this information to exist?) again leads you to consider what people and organizations may have a financial stake in a problem. Asking "Who would pay to know about auto emissions?" does not lead you to a library but to the names of key lobbies, government research bureaus, and manufacturers.

Research Strategies

You may ask how you can apply this advice. The answer is that until you gradually develop expertise in bibliographic or library searching, you will start your research with a person—either a reference librarian or a faculty member who can steer you in the right direction. Then your search will include at least three co-ordinated thrusts:

1. Searching for current information in libraries and checking reviews to make sure that your data come from respected sources.
2. Contacting experts in person or by phone or letter; contacting organizations that focus their efforts in your area of interest.
3. Using computerized data base searches and library networks to check older information in print as well as recent publications.

If you are not familiar with the indexes, abstracts, organizations, and data bases in your field, consult a full-length technical writing text or start by asking your local librarian to introduce you to one or two key reference tools. Learning to research is like learning anything else—if you do it carefully and step by step, you will develop useful skills. Just as there is no quick way to learn statics or physical chemistry, there is no fast route to becoming an effective researcher or interviewer. On the other hand, you will probably find, as we have found, that by contacting experts you will gather excellent information more quickly than you anticipate. When you have collected your data and tested your hypothesis to the best of your ability, you are ready to organize a written report. Chapter 2 explains—step by step— how to organize and write a case report. The cases in this text, however, ask you to develop skills in two further areas of technical communication: giving oral reports and designing simple visual aids or illustrations.

ORAL PRESENTATIONS

Most jobs today require oral presentations that range from informal discussions of projects or products to formal presentations at professional conferences. There are many advantages to oral presentations: They provide you with the opportunity to get immediate reactions to your ideas from your audience, they allow you to present your material in exactly the order and with exactly the emphasis you want (as opposed to written reports where the audience can skip from section to section), and they are frequently easier for an audience to understand than a written report. However, there are also some disadvantages. Oral presentations are not as effective as written ones in providing information for later reference. In addition, there are limits on the size of the audience for oral presentations, and the audience may not react well to having the order of presentation decided for them by the presenter. However, on balance oral presentations are frequently an attractive choice, both in place of and in combination with written reports.

Preparing Your Presentation

The steps required in preparing an oral presentation are similar to those used in preparing a written presentation: Analyze your audience, define your purposes, and organize your data. Along with these steps, however, you need to plan the visual aids necessary to get your ideas across.

Step 1: Analyze Your Audience. Your first step in preparing an oral presentation is to analyze your audience in terms of their knowledge, interests, attitudes, and ability to act. You should determine first how much your audience knows about your subject and, consequently, how much they need to know. You should also consider their interest in the subject and any attitudes, positive and negative, they may have about it (and about you as the presenter). Finally, you should consider whether this particular audience has the ability to act on the information you are presenting. If they lack the ability to act, you may want to consider presenting your information to another audience—one that can act.

Step 2: Define Your Purposes. After analyzing your audience, your next step should be to define your purposes: What do you want that audience to do, based on the information you present? Oral presentations are usually either informative or persuasive; however, the line between these two purposes is not clearcut. A presentation designed to inform an audience about a new product may also be intended to persuade them to purchase the product at a later date. Similarly, a presentation designed to persuade an audience to implement a program would probably, of necessity, include a great deal of information about that program.

Step 3: Organize Your Data. You should organize your data based on the needs of your audience and your purposes in making the presentation. You should also consider your time limit. Most presentations last between 20 and 60 minutes.

The *beginning* of your presentation should present the necessary basic information, such as your subject and purposes, your scope (the material you will cover), and your goal—what you want the audience to do. You should also include some type of attention-getter—a surprising statistic, an anecdote about your subject, or even a joke if one is appropriate.

The *middle* of your presentation should present the major points along with the most important supporting data. Don't try to cram too much information into your presentation; remember that people have limits on the amount of data they can absorb at one sitting. Order your points according to your audience's needs and your own purposes; usually the most important point, from the audience's point of view, is presented first.

The *ending* of your presentation is your last chance to tie all of your points together. You should summarize your major points and present any recommendations you have to make. Close your presentation with a "hook"—a final point (e.g., the next step that should be taken or what may happen if action is *not* taken) that will get your idea across and indicate to the audience that your presentation is finished.

Throughout your presentation you should use *transitions* to tie your material together. Indicate when you have finished one point and are moving on to another; for example, "That's the design for the old frimbus. Now let's consider the new, improved model." In longer presentations you can also include minisummaries to summarize each section before you move on to the next one.

Step 4: Plan Your Visual Aids. Visual aids are an important part of any oral presentation; they distinguish an oral presentation from a speech, in which the dynamism of the speaker is the sole means of maintaining the audience's interest.

You should plan your visual aids as you plan the spoken part of the presentation; audio and visual should be integrated thoughtfully into the whole; both are equally important. There are two common techniques for planning audio and visuals together. With a *script* you divide a piece of paper in half vertically. On one side you outline the spoken part of the presentation; on the other side you sketch the visuals that will accompany each point in the audio section. *Storyboards* can be used in a similar manner; using storyboards you place segments of the audio and a rough sketch of each visual on notecards.

Visuals can consist of pictures, graphs, charts, or simply words (e.g., "agenda slides" that list the major points of the presentation in order). They can be projected (slides, overhead transparencies, videotapes, films) or nonprojected (posters, flipcharts, models, demonstrations). As a general rule nonprojected visuals are used only for small, informal presentations. Projected visuals are used with larger groups.

Follow these seven rules when designing visuals for your oral presentations:

1. Keep your visuals simple, only one point per visual.
2. Avoid unnecessary words and details.
3. Make your visuals readable; use large type and lots of white space.
4. Don't simply take your visuals directly from a written report; redesign and simplify them.
5. Put information at the top of the visual and orient your material horizontally so that those in the back of the room can see it.
6. Use a consistent type and art style in all of your visuals.
7. Make your visuals of high quality; the quality of your visuals will reflect on you as presenter.

Delivery

When you deliver an oral presentation one of your goals is to come across as the best possible version of yourself. Address the group as if you were in normal conversation with them, avoiding stiffness and unnecessary formality, but also avoiding inappropriate casualness. Speak articulately and with animation, avoiding a monotone.

Avoid physical and verbal distractions; listeners are annoyed by rattling keys or a repeated *you know* and may stop listening to the points you are trying to make. Try to make eye contact with your audience throughout your presentation; do not stare down at your notes or behind you at your visuals. It's best to deliver your presentation from notes, an outline, or cues within your visuals; it's seldom necessary to *read* a presentation. Reading is the least effective technique from an audience's point of view. When referring to your visuals be careful not to block the view of anyone in the audience.

Finally, as you deliver your presentation, be aware of your audience and its reactions. Speed up or skip some minor points if the audience seems bored or restless; on the other hand, slow down and present more details if the audience seems perplexed. Oral presentations give you a chance to judge the reaction of your audience directly, and you can take advantage of this fact to enhance your presentation.

VISUAL AIDS

As we have already said, reports and oral presentations often need visual support, especially when you are trying to convey information that is difficult to express in words. Spatial relationships, such as the converging lines of graphs or the position or shape of a mechanical component, generally require visual presentation. In addition, material that includes many numbers may be difficult for an audience to comprehend without visuals such as tables or graphs. The following seven rules should help you to design effective visual aids and integrate them with the text of your report.

1. Do not let your visuals stand alone. They should complement the text, not take its place.
2. Tell the reader when to look at a visual and where to find it. A parenthetical remark (see Figure 2 on page 18) will do.

3. Place the visuals *after* the discussion to which they relate, not before. Ideally the visual will appear on the same page as the relevant discussion, but this is not always possible or economically feasible.
4. Use a title to remind the reader about the subject of the visual. Readers often wait until they finish a paragraph or a page before turning to an illustration; therefore, your readers may need to be reminded of the subject and purpose of each illustration. Visuals should also be numbered in sequence.
5. In general, use one illustration to communicate one idea. A single graph with five lines that illustrate five different relationships is usually less communicative than five more simple graphs, each making a single point.
6. Never include a figure unless it is vital to your discussion. Sometimes it is necessary to throw away photos and graphs that you thought important at an earlier stage of your report writing process.
7. Don't neglect to use visuals in order to save yourself time or effort. Visuals can greatly simplify your readers' efforts to understand your work.

Types of Visual Aids

There are many types of visual aids of varying degrees of complexity. In this section we give you some examples of standard visuals and a brief indication of what each visual does best.

Tables. Tables are the best choice for displaying numbers when you do not need to compare rates of change. However, tables do help your readers to make comparisons between total amounts, as in Table 1-1. Tables can also be used for words, as in Table 1-2. Table 1-1 shows numerical data, whereas Table 1-2 gives instructions.

Planet	Orbital Speed (mi/sec)	Rotation	Satellites	Diameter (mi)
Mercury	29.8	59 days	0	3,031
Venus	21.8	243 days	0	7,680
Earth	18.5	23 hr 56 min	1	7,921
Mars	15.0	24 hr 37 min	2	4,218
Jupiter	8.1	9 hr 50 min	16	88,700
Saturn	6.0	10 hr 8 min	17	74,940
Uranus	4.1	10 hr 46 min	5	32,190
Neptune	3.4	18 hr 12 min	2	30,760
Pluto	2.9	6 days 9 hr	1	3,600

Table 1-1. A table with numerical data

Step	Action	Result
1	Pull down the CREATE menu.	
2	Highlight "New File" with the cursor.	
3	Press RETURN.	A new file will open on the screen with the cursor in the upper left corner.

Table 1-2. A table with nonnumerical data

Graphs. Basic graphs include line graphs, bar graphs, and pie graphs. (Other more complex graphs, such as histograms and frequency polygons, are beyond the scope of this discussion.)

Line graphs, such as the one in Figure 1-2, are best for showing the rates of change in data. *Bar graphs*, as in Figure 1-3, allow

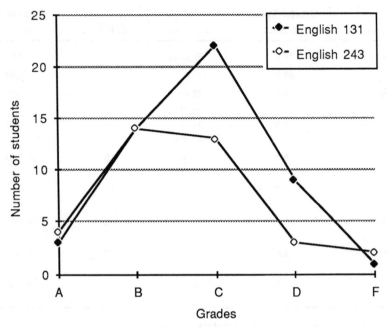

Figure 1-2. Line graph of English grades

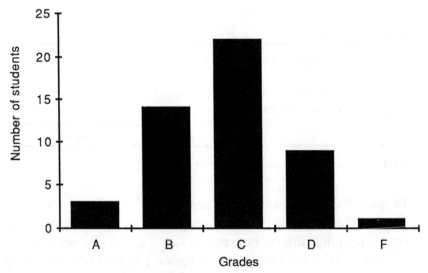

Figure 1-3. Bar graph of English grades

you to compare amounts more dramatically than tables. And *pie graphs*, such as the one in Figure 1-4, show amounts as parts of a whole.

Charts. The two most common types of charts are the organization chart and the flow chart. *Organization charts* show how agencies are organized. Figure 1-5 shows an organization chart for an academic division; notice that the chart moves from the top—the highest level of the organization—to the bottom—the lowest.

There are two types of *flow charts*: process and programming. The *process flow chart* diagrams the steps in a process. Figure 1-6 shows a process flow chart for making a transparency. Such charts frequently move from left to right, but they may also move from right to left, top to bottom, or be circular for a cyclical process. The *programming flow chart* diagrams a logic path and is most

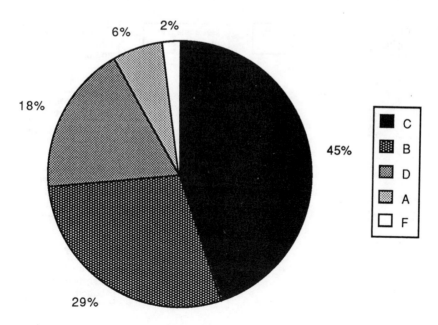

Figure 1-4. Pie graph of English grades

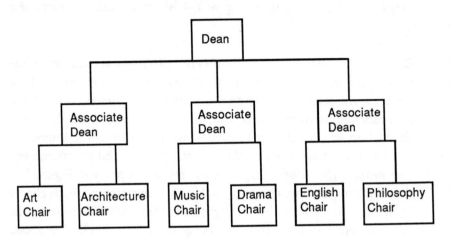

Figure 1-5. Organization chart

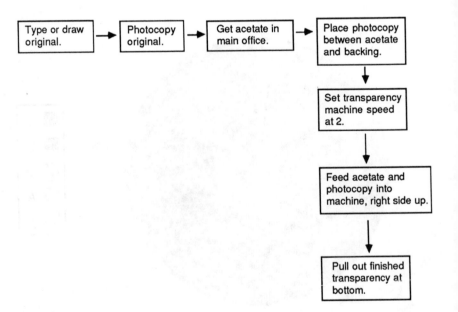

Figure 1-6. Process flow chart

frequently used in computer programming. These charts are arranged from top to bottom. The flow chart in Figure 1-7 diagrams the steps in organizing a paper.

SUMMARY

- Analyze your audience in terms of audience categories (general reader, technician, expert) and by completing an audience profile form.
- Assess the basic purposes of the report.
- Assign priorities and organize your work according to your priorities when you have multiple audiences and purposes.

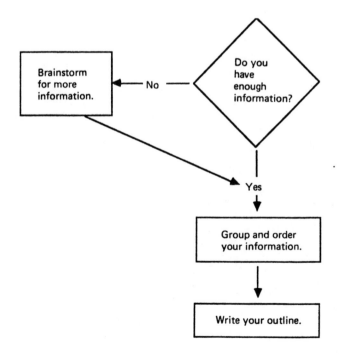

Figure 1-7. Programming flow chart

- Solve problems by brainstorming, developing criteria, clearly stating alternative solutions, and applying the criteria to the alternatives.
- Gather information by combining library research with focused interviews.
- Use oral presentations for immediate interaction with your audience.
- Use visual aids to support both written and oral presentations significantly.

HOW TO USE
THE CASES

INTRODUCTION

This chapter explains—step by step—how to prepare a writing assignment in response to a case. Reading this chapter will help you to organize your efforts. You will also gain a better understanding of the importance of the prewriting strategies of technical writing: audience analysis, the assessment of purpose, problem solving, and organizing and outlining the report. The Carlyle Windsor house case (from Trzyna and Batschelet, *Writing for the Technical Professions,* Wadsworth Publishing, 1987) presented here is a case assignment that calls for both a report and a letter.

Defining a Case

We begin by defining *case*. A case is a hypothetical situation that allows you to use your writing skills in a specific context. In the cases that follow you will be given a particular set of circumstances and, usually, a particular audience for which to write. You will also be given data to analyze and present.

Beginning on a Case

Your first job when writing a response to a case assignment is to analyze the situation you have been given. You must first define your audience and purpose and then select the data that will be relevant given that audience and purpose. If all the data are not specified in the case, you will have to find the necessary data through research. After you have defined audience and purpose and selected your data, you can organize and outline your material based on the audience's needs. Let's consider a sample case and see how this process works.

SAMPLE CASE: THE CARLYLE WINDSOR HOUSE

As your first job after receiving your B. S. in civil engineering you go to work for your friend Bill Morrasco, who owns Vintage Restorations. Bill's company specializes in restoring older houses in the historical section of the city, and they've been very successful.

For your first assignment, however, Bill asks you to clear up the company's first disaster. Six months ago, the company restored the Carlyle Windsor house, a large wood-frame Victorian on the city's west side. Extensive structural repairs were necessary, and these proceeded without problems. However, the final step was repainting the exterior. Something apparently went wrong; the new paint is now blistering and peeling. The owner of the house is understandably upset and demands an explanation, to say nothing of a new paint job. Bill asks you to investigate.

Your investigation produces the following facts (in no particular order):

The Carlyle Windsor house had had several coats of paint previously.

The choice of exterior paint was between latex and alkyd (oil base).

It had rained heavily two days before the house was painted.

Latex paints perform better than alkyd on damp surfaces because moisture passes through them readily.

The Carlyle Windsor house has a large elm tree on the north side; the south, west, and part of the east sides are in full sunlight.

Three days before the house was painted, the exterior was sanded down to bare wood, then a primer coat was applied.

Alkyd paints perform better than latex on previously painted surfaces.

Moisture meters are used before painting to test the percentage of moisture by weight of actual water present in the wood, plaster, or concrete.

Paint applied in direct sunlight may blister because the surface paint dries before thinner can escape; as the sun vaporizes the remaining thinner, the paint blisters because the thinner cannot escape through the outer skin of dried paint.

The percentage of water permitted in wood that is to be painted is usually 15 percent to 18 percent. 20 percent is too high.

It was partly cloudy on the day the house was painted; the high temperature was 76 degrees.

The primer coat should be allowed to dry at least two days before the finish coat is applied.

The moisture percentage found in the walls on the day the house was painted was 19 percent.

The forecast for the week after the house was painted called for clear skies and unusually high temperatures.

The project supervisor selected alkyd paint as the most suitable for the Carlyle Windsor house.

Allowing too long an interval between the application of the primer coat and the finish coat can result in intercoat peeling.

The most common cause of blistering and peeling is moisture.

Study these facts and arrive at a decision about the cause of the paint failure at the Carlyle Windsor house.

Assignment 1. Write a short report for Bill Morrasco in which you explain your findings and make recommendations.

Assignment 2. Write a letter to the owner of the Carlyle Windsor house, Dr. Lois Aragon, in which you explain why her house has a peeling exterior. Incidentally, Bill has decided to repaint.

WORKING WITH A CASE

Your first step in working with this case, as with any other case, will be to read it through carefully. Make note of the audience and purpose; begin thinking about how the material in the data section might be organized.

Analyzing Your Audience

After reading through the case your first step should be to analyze the audience. You have two audiences here: Bill Morrasco and

Dr. Lois Aragon. Consider, first of all, the knowledge of each of these audiences. How much of the information given in the case will be familiar and how much will have to be explained? In this instance your two audiences have very different degrees of knowledge. As owner of the company, Bill Morrasco should have a great deal of experience in the decisions and procedures involved in painting a house. You do not have to explain to him what a moisture meter is or why latex paint is sometimes chosen over alkyd. On the other hand, as the owner of a company that performs many different jobs, Bill may well have forgotten the exact circumstances of the Carlyle Windsor job, such as the fact that it rained heavily two days before the house was painted. Dr. Aragon, in contrast, probably has little experience with the requirements for painting a house. You probably need to explain to her the procedure used in painting as well as the probable cause for the peeling. Your audience's knowledge determines how much information you include and how sophisticated your explanations should be.

Next consider the probable interests of these two audiences. What is most important to them or, for that matter, least important? Bill is probably interested in exactly what happened and why; that, after all, is why he assigned the problem to you. Your conclusion about the cause of the problem and your recommendation (as well as a careful analysis of the circumstances) will probably meet his major concerns. Dr. Aragon is undoubtedly most concerned about what will be done to correct the problem. In her case the news that her house will be repainted is the most important, and it should be followed by some explanation of what happened. She is probably not interested in technical details, such as the exact percentage of water present in the wood, because she does not need to know them to understand the basic procedure the company followed. Include only the information that is relevant to your audiences' needs.

Defining Your Purpose

After you analyze your audience, your next step is to define the purposes of the report and the letter. You have a different purpose for each. For the report for Bill Morrasco your primary purpose is to recommend, while your secondary purpose is to inform. On the other hand, for the letter to Dr. Aragon, your only purpose is to inform. You can see how purpose affects your writing if you consider the changes that would be involved in your letter to Dr. Aragon if it were a recommendation that she pay to have the house repainted. In that case you would need to explain her alternatives and make a case for Vintage Restorations' refusal to repaint at its expense.

Organizing Your Data

The knowledge and interests of your audience will influence the way you organize your material, as well as the amount of material you include. You should always consider what your audience needs most when you consider how to order your data.

Your first step in organizing your data should be to *group* similar material. Look for common denominators between individual facts. You may discover that one group of facts on the list has to do with the choice between alkyd and latex paint. Another group of facts concerns moisture and its effect on the paint. Yet another group of facts concerns the amount of time between primer and finish coat, and a final group of facts concerns the effect of sunlight on paint. By grouping you can reduce this list of miscellaneous facts to four generalized groups and thus simplify the process of ordering your material.

Having grouped your data you can now proceed to *order* your groups, still considering your audiences and purposes. Because Bill Morrasco is most concerned with your conclusion and recommendation, these points would come first in your report to him.

Assuming that you concluded that moisture is the probable cause of the peeling, the facts about the rain and the moisture levels in the walls would come first, followed by an explanation of the choice of alkyd paint. Your recommendation may be to repaint the house, using latex paint if necessary. The other two points about the primer coat and the sunlight may be summarized at the end of the report, simply to indicate why you rejected them as possible causes of the peeling. In contrast, Dr. Aragon's main interest is the condition of her house; thus the information that the house will be repainted should be the first point in your letter to her, followed by an apology for her inconvenience. After that a brief explanation of the moisture problems will probably be sufficient. It is not necessary to outline the possibilities you rejected.

In this case you have only four major groups and a relatively small amount of data to order, thus the process is relatively simple. In other cases you may have many more groups and many more data. When cases are more complex you need to consider several factors to determine the most effective order: reader interest, logical connections between groups, even the relative importance of each group to your central point.

After all of this preliminary planning your final step is to write some kind of *outline* of your organization. For this case you may have already generated a simple outline in the course of your planning, and it may be sufficient for this particular assignment. However, in other assignments a definite outline will make your work much simpler; trying to juggle complex data without writing an outline in advance is more of a task than most of us can handle.

Planning Your Graphics

In some cases you may need to design graphics to support your discussion. They play a major part in most technical writing. Dealing with data of any kind usually requires some kind of visual support in the way of tables, graphs, drawings, diagrams, or charts. These visuals can be helpful both to your reader and to you; they make it easier for your reader to understand your discussion, and they help you to see relationships among your data.

In the case of the Carlyle Windsor house you probably don't need graphics to illustrate your points because the material is relatively simple. However, keep in mind that cases often require several visuals.

Having completed all of these preliminary steps, you can now proceed to write the report and the letter that were assigned in the case. They may look something like the report in Figure 2-1 and the letter in Figure 2-2.

SUMMARY

In working with a case assignment you should take the following preliminary steps before writing the final report:

- Read through the case carefully, noting audience and purpose.
- Analyze your audience.
- Define your purpose.
- Analyze your data.
- Group your data.
- Order your groups.
- Decide on introductory and concluding material.
- Construct an outline.
- Plan your graphics.

We need to stress one other point in discussing how to handle a case assignment. There is always a temptation, when dealing with a case, simply to regurgitate the case itself when you write your paper; you may find yourself copying the data word for word as

```
DATE:   June 4, 1987
TO:   Bill Morrasco
FROM:   Joanne McPhee
SUBJECT:   Problems with Paint at the Carlyle
            Windsor House


I have completed my study of the problems with
peeling paint at the Carlyle Windsor house,
825 Gardenia.  I concluded that the cause of
the problem was excessive moisture in the
walls, coupled with alkyd paint.  I recommend
that we repaint the exterior using latex paint
if moisture readings are high.


As you may remember, we began painting the
house on May 21 of this year.  The project
supervisor (G. Wilke) selected alkyd paint
because the house had already had many coats
of paint.  However, two days before the house
was painted, there were heavy rains.  The
moisture percentage in the walls on the day of
painting was 19 percent--below the 20 percent
cutoff but above the preferred 15 percent to
18 percent range.  The supervisor elected to
go ahead with the job, possibly because un-
usually high temperatures were forecast and
the house has little shade.  The high moisture
content was probably responsible for the peel-
ing.
```

Figure 2-1. Sample report for Carlyle Windsor house case

I did investigate the possibilities of intercoat peeling or excessive sunlight; however, I
do not feel that intercoat peeling is a probable cause in this case because the primer
coat had been on the walls for only three
days, two of them rainy. By the same token,
although the house is in full sunlight on the
south, west, and part of the east sides, the
day the paint was applied was partly cloudy
and relatively cool.

Once again, I feel that moisture was the problem. If the weather is more favorable this
time around, we should have no trouble.

```
Joanne McPhee
Construction Engineer
Vintage Restorations
110 Sprucewood
San Carlos, TX 78311


November 23, 1988


Dr. Lois Aragon
825 Gardenia
San Carlos, TX  78311

Dear Dr. Aragon:

On behalf of Vintage Restorations, let me
apologize for the problems you have had with
peeling and blistering paint.  We are ready to
strip the exterior walls of your home and
repaint whenever the weather permits; we will
contact you to discuss convenient dates.

Mr. Bill Morrasco, owner of Vintage Restora-
tions, asked me to investigate the paint prob-
lems at your home.  I have concluded that the
heavy rains two days before the walls were
painted resulted in a high moisture level in
the wood.  Although we do measure moisture in
```

Figure 2-2. Sample letter for the Carlyle Windsor house case

the surfaces to be painted using a device
called a moisture meter, the reading for the
walls on the day of painting was on the
borderline between being acceptable and unacceptable. Our supervisor elected to proceed
with an oil-based paint, normally a good
choice for a house like yours. Unfortunately
the combination of moisture and oil-based
paint resulted in peeling.

Again, let me assure you that Vintage Restorations stands by its work. We will repaint the
Carlyle Windsor house, being careful to allow
for excess moisture and to choose the correct
paint.

Please contact me if you have any questions.

Sincerely,

Joanne McPhee

part of your discussion. However, this approach does not produce a very effective paper. Usually the data in the case have not been appropriately organized; frequently they are presented in phrases that are neither parallel in construction nor complete sentences, and there are sometimes other grammatical errors. Finally, copying someone else's material is plagiarism even if it is inadvertent.

All students in your class will have access to the same material in a case assignment; the challenge is to interpret, analyze, and synthesize that material in your own way to present the most effective version you can.

BUSINESS COMMUNICATION

THE FOUR BASIC PRINCIPLES OF BUSINESS COMMUNICATION

You need to remember four basic principles of business communication when you are writing letters or memos.

1. *Be responsible.* Anything you sign is your responsibility. To the reader, you are the company. Write your correspondence intelligently and proofread it carefully.
2. *Address the reader personally.* Use "you" and "your" frequently. If you do so, your audience will read more rapidly, pay more attention, and be sympathetic to your message.
3. *Use simple, conversational language.* People like straightforward talk. Stay away from the cliches of business writing, such as "it has come to my attention that. . . ."
4. *Use positive language.* You can show that you trust your readers by expressing all of your news—even bad news—in positive language. Negative statements, such as "we are sorry that" or "unfortunately" or "we regret to inform you," irritate readers.

Format

The most common letter format used in business is the semi-block format. You do not indent the first line of each paragraph. You use single space within paragraphs and double space between paragraphs. Everything is justified on the left margin except the date, the complimentary close, and your signature. Generally you type the date two lines below the company letterhead. If you do not know the name of the addressee, write to a department rather than using awkward forms, such as "to whom it may concern" or "Dear Sir or Madam." It is preferable to write "Dear Credit Department" or "Attention: Credit Department." Close your letter with "Sincerely" or "Sincerely yours" unless the situation clearly calls for a more cordial or more formal close, such as "Cordially yours" on the one hand or "Respectfully" on the other. When a letter extends to more than one page, the full name of the recipient, the page number, and the date should appear at the top of each page. For further details, consult a handbook.

TYPES OF LETTERS

The remainder of this section discusses the memo and four common types of business letters: persuasive letters, bad news letters, complaint letters, and direct request letters.

Memos

A memo (or memorandum) is a short, usually informal internal report; memorandum means reminder. Memos should normally be kept under two pages. When you write a memo, provide the context rather than assuming that your reader remembers it. Ask yourself all the standard journalist's questions: who, what, when, where, and why. Memos may be less formal than correspondence to people outside the company. Often memos use the headings shown in Figure 3-1, with the writer's signature or initials written next to the writer's name on the "FROM" line.

Persuasive Letters

Most advertising follows the persuasive letter format, which you should use whenever you want to sell an idea. This type of letter has four parts: (1) an opening that attracts attention, (2) a section that develops the reader's interest, (3) a clincher that proves your main point, and (4) a specific action the reader can take. Advertisers generally attract your attention by clever graphics, promises, predictions, offers, or testimonials. The same types of material are used in the development section, which is sometimes called a *chain* because it is a string of persuasive facts or assertions. In Figure 3-2 the third part—the proof—consists of a testimonial and a statistic, while the final part mentions a return mail card and a toll-free telephone number. These principles work well in any situation, and they are easy to learn.

Bad News Letters

Bad news letters come in two forms, the direct and the indirect. Use the *direct* bad news letter when the situation is either minor, as it is when there is a slight delay in mailing an order, or urgent,

```
TO:   Sam Dunne
FROM:  Joanna Roberts
DATE:  October 15, 1987
RE:  Administrative Personnel Questionnaire
```

Figure 3-1. Sample memo headings

INTERPART INCORPORATED
P. O. Box 45
Bellevue, Washington 98456
1 (800) 368-3639
CABLE INTERCORP

April 30, 1987

Ms. Roberta Williams
Director of Purchasing
Expert Engineering
Worthington Parkway 68
Industry, California   91040

Dear Ms. Williams:

[Attract Attention]
The Interpart computerized on-line ordering
system can save your purchasing department up to
20 percent of its operating costs and cut a week
off the average time it takes to fill a parts
order.

[Develop]
Interpart is a new on-line computer network that
links 160 major industrial suppliers of equipment and materials. By using the Interpart
system, you can compare suppliers' prices and be
instantly aware of special purchase opportuni-

Figure 3-2. Sample persuasive letter

ties. You pay only for the system time you actually use for business transactions, and you can list your own products and surplus materials for a nominal fee.

[Prove]
More than 50 major industrial and engineering firms have become Interpart subscribers in the last six months. One user writes: "Interpart has enabled us to expedite rush orders and to take on contracts we would have had to pass by."

[Specify Action]
If you would like more information about the services and materials available through Interpart, as well as a full list of subscribers, please fill out the enclosed card or call 1 (800) 368-3639. We will respond immediately. Interpart--the industrial purchasing system of the future.

Sincerely,

Ellis Plantinga
Director of Marketing

Enclosure: Interpart Action Card

as it is when a bill is long overdue. The direct bad news letter has three short sections: (1) an immediate statement of the bad news, (2) a short explanation, and (3) a courteous close that includes a sentence to encourage more business. In bad news letters, you should always avoid negative language and accusations. Figure 3-3 exemplifies the characteristics of a bad news letter written in response to a minor problem.

An urgent, direct bad news letter might be the last communication in a series of collection letters (see Figure 3-4).

The *indirect* bad news letter has four sections. You start with a *buffer,* which is a positive and encouraging statement that places your bad news in context. You can thank the reader for something, compliment past service or promptness, or make some other positive remark. Second, give an *explanation* of the situation without blaming anyone. State the facts and the policies that apply. Third, state an *implied rejection,* letting your reader gradually understand that the news is bad. Fourth, use a *positive close* that affirms your belief in the reader's fairness, intelligence, and understanding. The rejection letter in Figure 3-5 displays these principles in action.

Before you send a bad news letter, read it over to be sure that you show trust in the reader's fairness. Also check for negative language. If you must imply blame or responsibility, blame an individual as a member of a group. In other words, don't say, "Mr. Smith, you should have understood . . . ," say, "Some people don't understand that. . . ."

Complaint Letters

Complaint letters combine features of bad news letters and persuasive letters. You need to assure the reader of your fairness and your general past satisfaction with the company's service or product. You must assume fairness on the part of the recipient, and you should state all the relevant facts clearly. As in a persuasive

Dear Ms. Lockwood:

The proofs are good, but there is a missing
page (17) in Chapter 3. When you find this
page, please send it to me for proofreading.
I also found and marked three minor errors
that need correction.

Sincerely,

Anna Trompkin

Enclosure: page proofs, set 3

Figure 3-3. Sample direct bad news letter for a minor problem

Dear Mr. Smith:

Your overdue account has been turned over to
the Ajax Collection Agency. If you wish to
avoid legal action, mail full payment by the
end of this month.

Mail your payment to:

Ajax Collections
P. O. Box 1489
San Diego, CA   92999

Sincerely,

Alan Smith
Credit Officer

Figure 3-4. Sample direct bad news letter for an urgent problem

Dear Ms. Kalman:

[Buffer]
Thank you for applying for the position of
Engineering Intern. We are highly impressed
by your training and experience.

[Explanation of Criteria for Selection]
Seventy-five student engineers applied for the
two advertised openings, many of whom also had
fine records. As we considered the applications, we narrowed the field to a few students
who had taken advanced courses in Pascal, File
Structures, and machine programming, and who
had previous experience with manual writing.

[Implied Rejection]
We will keep your application on file in case
we have an opening that more nearly fits your
qualifications and experience.

[Positive close showing appreciation for the
reader]
Thank you again for applying. It was a plea-
sure to read such an excellent resume.

Figure 3-5. Sample indirect bad news letter

letter, specify the action you want taken: a refund, an adjustment of a bill, a new product, or whatever else you think is fair.

Direct Request Letters

Direct request letters have three parts: (1) the main request, (2) the explanation, and (3) the recapitulation and close. The main request should specify all the details of the request. If you have several requests or questions, they should be numbered and presented in separate paragraphs for easy reference. Word your questions in a neutral way. In the explanation paragraph, develop your own credibility and trustworthiness—your worthiness to get what you are asking for. You should also show how the reader will gain something by responding to your request. If you ask a company for information for a report, for example, you can indicate that the company will get advertising and good will by having its name mentioned prominently in the report.

CASES

PARK AND RIDE LOTS

As an assistant urban planner for the city of Santa Clara you have been involved for the last month and a half in a major reassessment of the city's traffic plans. As the number of commuters to the central city has increased, so have problems with snarled traffic, parking shortages, and auto accidents. All of you on the urban planning staff have been researching options for the city's traffic control committee (made up of three city council members and the city planner) to try to relieve the situation, and you think you have come up with a promising possibility: park and ride lots.

The idea behind the lots is to transfer commuters from their own cars to buses for the trip to the central business district. The commuters would drive to centralized locations (you are currently considering the parking lots of some outlying shopping malls, as well as some city-owned land near major highway interchanges), park their cars, and transfer to shuttle buses. The buses would then deliver them to several locations in the downtown area.

One way to judge park and ride lots is to analyze their benefits versus their costs. The following are some of the benefits you have found for park and ride lots:

- Reduction for consumers in the cost of owning and operating a vehicle (fuel, repairs, tires, insurance, depreciation, etc.).
- Reduction for consumers in parking costs.
- Reduced miles of travel for vehicles.
- Reduced energy consumption.
- Reduced vehicle emissions.
- Reduced traffic congestion.
- Reduced parking demand in downtown areas.
- Reduced nonquantifiable consumer costs (for example, reduced stress, increased safety, companionship, and so on).

You have also found costs for both capital investment and annual maintenance and operation. Averaging the costs for the cities already using park and ride lots, you have arrived at the following costs:

- Capital investment: $1000 per parking space
- Project life: 5 years
- Salvage value: 0
- Interest rate: 15 percent
- Operation and maintenance per year: $40 per parking space

The capital costs include lighting, signing, marking, and other features needed to make facilities operational.

In your own state you have found six cities operating a total of thirty-five lots statewide. Three other cities are considering building such lots, and the original six are all in the process of adding more lots to the total. The reasons you have been given for using these lots include the following:

- Parking costs in the downtown areas are increasing and available parking spaces are decreasing; three of the largest cities in the state have adopted policies prohibiting the building of new parking lots in an attempt to discourage auto traffic.
- Fewer roads are being built, and more emphasis is being placed on increasing the capacity of existing roads to carry more people.
- Costs of owning and operating a car are increasing.

You have also been given an opportunity to measure what commuters themselves think about park and ride lots. The Transportation Study Group, a private foundation, surveyed the commuters in the six cities that have adopted park and ride lots to discover how they reacted to the park and ride experience. The questionnaire the Transportation Study Group used is shown in Figure 3-6.

The survey had a 50 percent response rate overall for the six cities. The answers give you some interesting information.

On the question of time, 15 percent of those responding thought that they saved time. Forty-six percent felt that they did not save time, and 27 percent were not sure. The other 12 percent reported no change. The study group believes that the loss of time is caused by the time it takes to transfer from car to bus, which sometimes involves waiting for the bus. Those who lost time estimated they lost an average of 15 minutes each way.

On the question of money, the results were more favorable. Ninety-five percent of those responding felt that they saved money by using the park and ride lots. Two percent felt that they did not save money, while 3 percent were uncertain. The average monthly savings was $60. The average daily cost for the park and

Transportation Study Group
Park and Ride Questionnaire

1. Before using Park and Ride, how did you
   make this trip?
   __ Personal Car __ Carpool __ Bus
   __ Other (please describe):

2. How long have you used Park and Ride?

   _____

3. Do you save time using Park and Ride
   rather than driving yourself?
   __ Yes. How many minutes do you save one
   way? _____
   __ No. How many minutes do you lose one
   way? _____
   __ Not sure
   __ No change

4. Do you save money using Park and Ride?
   __ Yes. How much do you save per
   month? _____
   __ No. How much do you lose per
   month? _____
   __ Not sure
   __ No change

Figure 3-6. Questionnaire for park and ride lot case

5. How satisfied are you with Park and Ride overall?
   __ Very satisfied __ Satisfied
   __ Neutral __ Dissatisfied
   __ Very Dissatisfied

6. How could Park and Ride be improved?

Thank you for your cooperation.

ride users was $2.70 for both fare and parking fees at the park and ride lots; the study group estimated that park and ride users could save about $700 annually.

In the largest city in the state the total construction cost for nine park and ride lots was $970,000; maintenance cost $39,000 annually. The annualized cost was $291,000. The city found that 955 commuters used the facilities, reducing the amount of miles traveled annually by 6,185,000. The annual savings in operating costs was $795,000, and the annual savings in gallons of fuel was 380,000.

Assignment 1. Now that you have all this miscellaneous information, you must put it into shape for the members of the traffic control committee. You decide to write a memo outlining your findings. Your primary purpose here is to pass on the information you have learned about park and ride lots. Consider what information will be of most interest to your readers and what order would best convey that information.

Assignment 2. The general reaction to your memo was favorable. The committee has decided to recommend the park and ride option on a limited basis. You have been drafted to write a recommendation report in the form of a persuasive letter to the city council, suggesting the construction of three park and ride lots. Each lot would have approximately one hundred spaces.

THE RUBBER FORMULA

As head of the maintenance laboratory at Eastland Air Force Base, your job involves many duties, including quality control and environmental testing. However, you also fulfill requests from the various departments at Eastland, and you have just received one from Arvel Harris of the aircraft maintenance department. His request follows:

> We've been trying to mold this rubber window gasket for about 5 days now. It's about 18 inches across and there are 10 hollow steel pins, ¼ in. in diameter, that hold it in place in the mold. We've been using Universal Latex's SR4 rubber because it's got properties we wanted in the part—it's hard enough and has good elasticity. The problem is, as the rubber sets up, it snaps those little pins; it gets *too* hard, in other words. When that happens the gasket deforms; it's ruined. What we need is to have you analyze the SR4 and tell us what's in it so we can maybe change the formula enough to make it work for us.

This request is not unusual, but in your experience it's not really one you can fulfill. Your analysis can usually tell you what kind of rubber is being used in general terms, but it can't tell you precisely what substances have been added to the basic rubber type to give it whatever unique properties it has. After a preliminary test on the SR4, your suspicions are confirmed: You cannot provide Harris with the exact formula he needs.

Your next step is to write to Universal Latex, the manufacturer, for help. What you want is the SR4 formula so that you can adapt the rubber for Harris's purpose.

Assignment 1. Write a memo to Arvel Harris, explaining the problem with your analysis and telling him what you plan to do next.

Assignment 2. Write a letter to Judith Binns, head of Universal Latex's product development laboratory, asking her for the formula for SR4.

At Universal Latex, Judith Binns has a different problem. The formula for SR4 is considered "proprietary information"; Universal does not want to supply the formula to anyone outside the company because the formula might fall into the hands of Universal's competitors. Thus Binns has to refuse the request from East-land Air Force Base.

However, Eastland is a major customer for Universal Latex products; Binns definitely doesn't want to offend the head of the laboratory or the head of the aircraft maintenance division. More-over, they obviously have a problem with SR4 in this application. After considering the information supplied in the letter from the laboratory chief, Binns feels that another Universal product, SR11, may meet their needs. Thus she decides to suggest they use SR11 instead of SR4.

Assignment 3. Write the reply Judith Binns sends to the head of the laboratory at Eastland Air Force Base.

THE GAS CHROMATOGRAPH

For the past three years your lab has been struggling along with its old gas chromatograph. The instrument was there when you first came to the lab; it was somewhat outmoded then but still compar-able to what other labs were using. As you became aware of the new instruments available, you were more and more convinced that your chromatograph should be replaced. Your group leader was aware of the problem, but other equipment purchases took priority. Besides, the old chromatograph was still useable for most of the lab's applications. Then six months ago your group leader decided the time was right to order a new, state-of-the-art instrument.

The new instrument has all the features you've been consider-ing: microprocessors that control the oven temperature, the key-board and monitor, and the data output. Unfortunately, it took you a while to discover the system's capabilities. The installation required expert supervision, and the local service representative for Standard Instruments (the manufacturer of the chromatograph) was always too busy to come to the lab to oversee the job. You tried repeatedly to reach him—three times on June 30 (the day after the instrument was delivered), twice on July 1, three times on July 2, and five times on July 3. You left a message each time, but it wasn't until the afternoon of July 3 that the service representative finally returned your calls. Then he explained that he would be unable to come to the lab until July 8 because of a backlog of work caused by the July 4 holiday. You weren't pleased, but you agreed. What else could you do?

When the service representative finally arrived, he was unable to install the instrument because the capillary column was missing. As the service rep explained it, the column had to be purchased separately because there were several column options. You informed him that you had ordered the capillary column when you ordered the chromatograph; it was your understanding that the column would be shipped with the instrument. However, the column was undeniably missing, and the service rep promised to call the main office of Standard Instruments and order the capillary column.

This time you had to wait six weeks. You kept in contact with the service rep; at least you left messages with his office every three to four days. He seldom returned your calls. At the end of six weeks he called to inform you that the capillary column had arrived and that he would come to the lab the next day to install it. Now it was your turn to cause a delay; the following day you were planning to leave to attend a professional conference in Chicago, and you were going to be gone for ten days. The service rep could have installed the column in your absence, but you wanted to be present when the tests were run so that you could see first hand how it was operated. You asked the service rep to reschedule the installation for two weeks later; reluctantly, he agreed.

When you returned from Chicago the same problems resumed. The service rep broke the appointment you had made to have the instrument installed because of an emergency service call from another city. When he tried to reschedule the appointment, he picked a time when you were once again going to be absent from the lab. Finally, a month later, your two schedules coincided. The instrument was installed and you learned how to use it. For the next two weeks, the chromatograph worked well; you were thoroughly satisfied with the instrument if not the local service representative.

Then problems arose with the chromatograph's oven. One day as you tried to heat your column in the oven, it became obvious that there were serious problems somewhere in the system. Again, you called the service representative; again, you were forced to wait for a week before your calls were returned and another two weeks before the rep could make it to the lab. His examination of the oven confirmed your fears: The heating coil was cracked and the oven was inoperable. You had always used the oven according to the directions in the instrument manual; the service rep agreed that the part itself was probably defective.

You resigned yourself to another extended wait, but this time the news was even worse than you had anticipated. When you tried to arrange for the necessary replacement, the service representative told you that the chromatograph had only a four month warranty—four months from the date of purchase, not the date of installation. The delays had effectively eaten up the warranty period before the instrument was even in use. The oven was obviously defective; it is obviously not your fault that the heating coil cracked. Nonetheless, according to the service representative, you need to pay full price for the replacement.

Assignment 1. Write a letter to the customer service representative at Standard Instruments, Ms. Dora Chung. Your purpose is to get the company to replace the cracked heating coil under the warranty.

Assignment 2. The director of your division, Dr. Howard Crossman, has been asking why the gas chromatograph is not in service. He has tests he wants to run. Write a memo to Dr. Crossman explaining what happened.

Assignment 3. Assume you are Ms. Dora Chung, the customer service representative at Standard Instruments. Write a reply to the customer who has written to you to complain about the cracked heating coil and the poor service received from the service representative.

THE FLUORIDE REFERENDUM

After graduating from dental school you returned to your hometown to set up your practice. Everything went well except for one frustrating experience. About a year after your return there was a local referendum about fluoridating the town's water supply. The dental society, of which you are vice president, and other local groups were solidly in favor of the measure, and you all assumed that the benefits of fluoridated water were so well known that the referendum would pass easily. To your surprise the referendum was defeated by a narrow margin.

After the election was over the dental society discovered that a small but well financed antifluoride campaign had been carried on in several of the more populous suburbs of the city. The campaign consisted of several anonymous pamphlets featuring wildly inaccurate statements about the "dangers" of fluoride. After doing some digging you discovered that the campaign was financed by a local eccentric, Otis Burnside. Although Burnside paid for and probably wrote the pamphlets, he never made any public statements about fluoride that could be refuted.

Now, however, that may change. The city has decided to resubmit the fluoridation referendum because the state has just made funds available for fluoridation. Today the following piece appeared in your local newspaper's Sound-Off column, which prints essays submitted by readers.

It's Your Turn:
The Fluoride Trap

Once again the city is trying to push down our throats the nefarious plans of the professional fluoridators. Amid all this talk of the so-called "benefits" of fluoride there are several facts they never seem to mention.

- Fact #1. Fluoridating our water supply would be expensive. Not only would the city be forced to pay for the fluoride to add to our water, they would also have to spend millions on the equipment needed to add this poison.
- Fact #2. The "benefits" of fluoride are far outweighed by its *proven dangers.* Scientific studies have proved that fluoride can cause cancer, birth defects, and other serious health hazards. Measure this against the questionable effects fluoride has upon tooth decay!
- Fact #3. Fluoride is a *deadly poison.* One teaspoonful of fluoride would be enough to kill the average person. Yet this is the substance your city government wants to add to the water you drink every day!

My fellow citizens, let those who wish to give this poison to their children do so privately. But keep this substance out of our water supply; our very lives may depend upon it!

      —Otis Burnside

Your first reaction after reading the column is anger: Even though the column appears on the editorial page, you're afraid many people will believe that Burnside's statements are actually true. You decide that the editor should in good faith print a rebuttal to Burnside's melange of half-truths and falsehoods.

Assignment 1. Write a letter to the editor of the newspaper asking for equal space to reply to Burnside's column. Remember to include specific reasons for making your request.

The editor agrees to give you space for a reply. He limits you to 1,000 words (approximately five typewritten pages), but he says that he will back up an appropriate response with an editorial supporting the fluoride referendum.

As you consider what to say in your reply you assemble the following facts:

Fluoride in all plants, soils, and fresh water supplies; considered an essential nutrient (National Research Council; National Academy of Sciences).

20,000 studies of fluoride since 1950; 3,700 since 1970.

20% of fluoride retained in mouth fluids and deposited on tooth surfaces; 80% enters the system. Most excreted. Of small amount remaining, 96% goes into bones and teeth. Soft tissues don't retain fluoride (U.S. Department of Health and Human Services).

Children absorb fluoride most rapidly; after age 8, absorption declines.

Fluoride strengthens tooth enamel; becomes part of tooth surfaces and some of the inner structure. Strengthened enamel can resist action of acids formed by oral bacteria. (U.S. Department of Health and Human Services)

Highest tooth decay activity found in school-age children from childhood to adolescence.

Almost 116 million Americans (50 percent of the total users) use fluoridated water. Approximately 59 percent of U.S. drinking water is optimally fluoridated (Texas Department of Health, Dental Bureau)

Optimum levels of fluoride dependent on average temperature of locale. Hotter the place, lower the amount of fluoride needed—citizens of warm locales drink more water, absorb more fluoride. Our average daily high temperature = 63.9° – 70.6°. Optimum fluoride level 0.9 parts per million. Water supply already has 0.6 parts per million. (U.S. Department of Health and Human Services)

At optimum levels, fluoridated water reduces tooth decay in children from birth to 12 years old by 65 percent. It reduces tooth decay for children 10 to 16 years old by 40 percent to 50 percent. (Texas State Technical Institute)

Fluoridation would cost approximately 25 cents per person per year. Most dentists estimate that each filling has to be replaced approximately 5 times in a lifetime. A filling currently costs around $20. (Oral Health Association)

There has never been a clinically substantiated case of disease or other harm caused by the consumption of optimally fluoridated water. (Texas Department of Health)

Only substantiated risk from fluoride: mottling (discoloration) of teeth. Occurs only at 2 to 8 times the optimal fluoride level. (U.S. Department of Health and Human Services)

At levels in excess of 8 parts per million, continued fluoride consumption over a number of years can result in calcification of ligands and tendons (osteosclerosis). (U.S. Department of Health and Human Services)

The toxic dosage of fluoride is 2 grams. For a toxic dose in a 10-ounce glass of water, you would have to add about 700 tons of sodium fluoride to the water supply at one time. The usual amount added is 50 pounds every two days. (U.S. Department of Health and Human Services)

Fluoride can be applied by a dentist or taken in tablet form. These alternatives are only 1/10 as effective as fluoridated water, and they cost 9 to 10 times more. (U.S. Department of Health and Human Services)

Assignment 2. Write a column refuting Burnside's antifluoride statements. Your audience is the general public, in particular those who will be voting on the referendum concerning fluoridation.

Assignment 3. A local citizens' group has invited you to speak about the fluoride question. Prepare a script (audio and visuals) for an oral presentation based on the data.

RAISING THE WATER RATES

Now that the city of Columbus has decided to raise its water and sewer rates, someone on the staff of the water department needs to take the formal discussion of the rate hike—a discussion prepared for county and city government officials—and turn it into two simple documents for other audiences. First, before the actual rate hike figures are announced, a bad news letter must be sent to all consumers, notifying them that in two months their utility bills will be going up substantially. Second, as part of a continuing study of water and sewer rate changes in comparable cities, someone must write a direct request letter to officials in those other cities asking for information about their studies and responses to the five or ten main issues related to sewer and water costs. The document in Figure 3-7 is the overview of the issues prepared for discussion by the city council and the county utility commissioners. You have been selected to write the two letters.

Assignment 1. Write a short bad news letter to the residential water consumers in Columbus. This letter will be enclosed with next month's water bill and will notify them that new rates will be announced in two months. Explain the reasons for the rate hike. Be sure to consider all the consumer groups and their concerns.

Assignment 2. Write a direct request letter that will be sent to the utility managers of comparable cities asking for information that will be of use to Columbus's Rate Hike Committee as it continues to investigate alternative rate structures and conservation measures.

```
        OVERVIEW:  WATER AND SEWER RATE HIKE

The City of Columbus Water Treatment Authority
owns and maintains a sewer and water system
that serves more than 170,000 customers.  Af-
ter maintaining level sewer rates for more
than ten years, the city needs to revise its
rates to cover mounting costs of maintaining
and operating the system.

Cost increases can be traced to several
causes:  inflation in the costs of maintaining
and operating the system, including the cost
of chemicals; a need to replace aging equip-
ment or to rehabilitate or extend sewers and
treatment plants; added expenses because of
new federal standards for waste treatment and
effluents; and increased capacity due to ser-
vices offered to other districts.  However,
the city´s population has not grown at the
same rate as the costs, so that there is no
growth in sales to help carry the added costs.
Consequently, the costs must be passed on to
the nearly stable pool of consumers in the
form of higher rates.

During this same period, utility·rate deci-
sions have drawn increased attention from both
national and regional government officials and
```

Figure 3-7. Overview of issues involved in the water and sewer rate hike

increased scrutiny by citizen's organizations, environmental advocates, and concerned customers. Technical details of rate decisions are of increasing interest to consumers, as are the methodological bases of rate scales, the policies that underlie rates, and the implications of rate structures, especially as they concern issues of reasonableness, fairness, human need, and environmental safety. Public participation in utility rate decisions has become more common, and, on the national level, many policy analysis organizations have studied this issue and made recommendations.

Altogether, five issues are central to the analysis of rates and the rationalization of restructuring utility rates. Careful financial management of the facilities is paramount. Economical use of resources is also important, as is the protection of the natural and human environment. The use of utility rates as part of programs to achieve large social or economic policy objectives also plays a role in decisions regarding the distribution of utility costs to various types and categories of consumers. Finally, general equity to consumers is an overriding goal that works hand in hand with the attempt to build a

rate structure that supports specific economic
and social policies.

In addition to the rising costs already
listed, the utility has also been faced with
the rising cost of the capital needed to upgrade and maintain facilities. Responding to
these incentives and challenges, the elected
officials responsible for Columbus's water and
sewer systems have attempted over the last
three years to institute programs that will
minimize costs and increases in operating
expenses. Two general approaches have been
applied. First, the utility has encouraged
conservation by mounting a campaign using
media spots, mailers enclosed with bills, and
school programs, and by providing inexpensive
water saving devices free of charge. Approximately 23 percent of homes in the Columbus
area now use shower head restrictors that
reduce the amount of water used in a shower by
50 percent. This direct conservation appeal
has been reinforced by a second mode of conservation: indirect incentives by graduated
rates. Those who use more pay more for the
additional increments.

Review by federal and state environmental
agencies has added an impetus to the conserva-

tion program. No rate revision and no capital
improvement plan is approved without a thor-
ough review of the potential for conservation
and the opportunity to produce an even more
environmentally desirable effluent.

At the same time, Columbus has been under
increasing pressure to assure that rate hikes
do not negatively affect two groups. Low
income consumers have been assisted by several
programs in the past, and the utility is seeking ways to assure that this group will not be
denied full use of city utilities because of
rising costs. Similarly, there are categories
of business and industry that must be pro-
tected against inflationary trends so that
they can maintain their competitive posture
with companies in surrounding areas, particularly some rural areas, where water and sewer
rates are still comparatively low.

As the city contemplates enacting a new rate
structure with substantially increased bill-
ings, the city council and county board must
consider how these changes will affect particular segments of the general population,
business, and light and heavy industry. It
may be necessary to introduce the new rate

structures gradually to avoid major disrup-
tions.

After a 15-month period of analysis, the rate
committee recommends that the rate structure
be changed next year to tie rates for sewers
directly to the volume of water used. At the
same time, the committee recommends the institution of a relatively high minimum charge to
all categories of consumers as the most effective means for leveling any abrupt changes in
the cost to particular consumers. The current
program of low income subsidies will remain in
effect. For the immediate future, the committee makes no recommendation concerning the
establishment of special categories of industries that may contract for rates that will
protect their viability and hence protect
their employees´ jobs.

CEDAR HEIGHTS

You have recently joined the professional staff of Urban Design Associates (UDA), a large urban planning corporation. Your immediate supervisor is the manager of the regional office of UDA. Urban Design Associates has just published the Draft Environmental Impact Statement for a new housing development in an old area of the city. For 40 years, a large block of land owned by the federal government has stood empty and surrounded by high, barbed wire fences. Finally, the federal authorities ceded the land to the city government, which decided to develop the property, conforming to several policies it adopted over the last decade. Since the property is across a large street from a major children's hospital, one acre will be developed by a popular restaurant chain which has committed a portion of its profits to building hospices for dying children and places where parents can stay inexpensively while their children are being treated. Another part of the tract is to be developed by charitable organizations that build housing for wheelchair-bound persons. The remainder of the property will be used for low and middle income housing.

The issue of neighborhood character was central to the impact study. The neighborhood behind the hospital consists of a small peninsula that extends into the lake that borders the eastern edge of the city. Social register data indicate that the citizens of this neighborhood, Cedar Heights, have one of the highest per capita incomes in the region. The other areas around the development are middle class to upper middle class.

In the environmental impact process, after a draft assessment of impacts has been prepared and published, the public has from 15 to 30 days to respond to the analysis with questions, comments, complaints, or challenges. Hearings are held, and the agency that produced the Draft Environmental Impact Statement (or their hired consultants) must respond to the comments. Today your office received a letter from the chair of the Cedar Heights Neighborhood Council. While this community leader has read the Draft EIS, he is particularly concerned about the measures that will be taken to mitigate any possible effects on the aesthetics and the property values of the neighborhoods surrounding the development area. Recognizing that the Cedar Heights Neighborhood Council will send representatives to the public hearings, that two city councilpersons live in Cedar Heights (as well as four state senators, not to mention many professors, corporate presidents, and long-time community activists), you know that it will be important to write a clear, well-documented, and persuasive letter to the council's chair. When you tell your boss about the mail, she asks you (of course) to write the persuasive letter. Reviewing the 120-page Draft Environmental Impact Statement, you make the following notes:

1. Site section drawings show building bulk, mass, and scale. Comparisons of proposed structures to current structures show no deviation from current patterns. (Hospital much larger, but that's another issue.)
2. Graphic overlays on photos of existing neighborhoods (based on 4-year-old aerial survey) show extensive visual, physical linkages between design of current neighborhoods and proposed structures, including proposed pedestrian/wheelchair bridge across major throughway to Children's Hospital.
3. Analysis of lighting patterns, total glare/lighting impact of new neighborhood acceptable.
4. Special attention paid to view corridors, vistas, and outlooks, especially from Cedar Heights, Children's Hospital, popular Murchison Bike and Running Trail that runs next

to property, and selected high ground in other surrounding neighborhoods.

5. Maintenance of property an issue in low income areas. Criteria for selection of developers and city property management plan (approved last year) demonstrate that sufficient resources are budgeted to maintain the new structures. City will own and manage properties, work will be done under contract by private corporations experienced in this field.

6. Impact on tax base/property values. Factors include mix of low income, middle income, high income properties. Racial issue? Data of study suggest race always less a factor than income. Stability of current neighborhoods most significant factor in evaluating this concern. (Hospital growth and traffic have had nil impact on property values in Cedar Heights, even though Children's Hospital has expanded into Cedar Heights land.) Property values affected by fear, self-fulfilling prophecies. Small, highly localized low income settlements historically have small impact. No potential for spread of low income area.

7. Nearest low income area? 4 miles? Mitigation? Will tax revenues from area cover cost of basic services (sewer lines, water, electricity extension, physical plant upgrades)? Probable. (Due to density of population anticipated.)

8. Mitigation? Plan includes extensive landscaping of main throughway. Bridge (pedistrian/wheelchair). Landscape median. Underground electrical service. New lighting system.

9. Crime? Increased lighting on throughway.

10. Plans include heavy new plantings of trees to replace loss of greenbelt. Previous government land = 15 acres of heavy growth, berries, pines, totally fenced.

11. Traffic analysis? Anticipated 50 percent new population serviced by van pool. No cars. Historically, traffic/pedestrian noise level near hospice: low.
12. Commercial analysis? No provision for extension of commercial zoning. Area will be served by single supermarket proximate to northern boundary of property.
13. How to address racial/social issue? Census 1980: predominately white area, with 2 percent black, 5 percent Asian, chiefly professionals with hospital, university, medical research center.
14. Development alternatives? Park? Park space included in plan—large park with ravine poses policing difficulties. sale to commercial developers for condo grouping? Population density high, traffic and noise a problem.
15. Probably see several council members at public hearing. Keep this one simple and basic. Go over details then.

Assignment 1. Write a persuasive letter to the Cedar Heights Neighborhood Council's chair to allay fears and offer cooperation.

Assignment 2. Your boss loves your letter. Now she wants you to prepare a handout sheet that can be distributed at the community meetings that are scheduled for discussion of the new development. She suggests the title "Development Alternatives for the Federal Property," and she asks you to present two types of information: (1) the city's alternatives before the choice was made to develop the multiuse hospice/handicapped/low and middle income area and (2) the features and advantages of the alternative the city chose.

TYPES OF REPORTS

This chapter discusses four common short report forms as well as recommendation reports and feasibility studies. The short report forms include the incident, progress, trip, and inspection reports. When you write a short report, you have a responsibility to report all the relevant facts and to interpret those facts for your readers. When short reports fail to communicate effectively, the problem is often caused by a failure to include important data or a failure to interpret the facts—to explain why they are being reported and what they should mean to the reader.

INCIDENT REPORTS

Incident reports are used to report information that can be fully analyzed and interpreted later. A typical incident report might be an insurance report on a fire or an accident. Be sure to cover *who, what, when, where, why, and how.* Incident reports are used to allow interested parties the time to consider and respond to the information before a final analysis or decision is made.

PROGRESS REPORTS

Progress reports are used to inform companies and the other people working on a project about the work you have completed, the problems and delays you have encountered, and the work you intend to finish next. Progress reports are also used to coordinate the activities of different work groups and to recommend or propose changes in schedules, equipment, staffing, budget, or design. Progress reports can be issued periodically or occasionally. In large engineering projects, working groups are often asked to file periodic reports (weekly, monthly, quarterly) that use a uniform format. When all the periodic progress reports are bound, they constitute a running record of everything that happened in the course of the project. Occasional progress reports, on the other hand, are written when there is a need for them—when some important bridge has been crossed or when circumstances call for a change in a project's direction.

The progress report outline differs from a basic three-part format (introduction, body, and recapitulation) because progress reports should link together to form a chronological sequence. The beginning of the report should mention the date of the last report, and the end of the report should indicate the date of the next anticipated report. The outline in Figure 4-1 suggests the major topics that should be covered.

TRIP REPORTS

Engineers and other professionals are often asked to write reports when they return from field trips, conferences, and other travels away from the office. Trip reports should include full information

```
1.  Overview
        Subjects and content to be covered
        Date of last report
        Main points

2.  Body
        Work completed in the period since the
            last report
        Work scheduled for the next period
        Problems encountered, solutions devel-
            oped, adjustments made or requested
            in schedule, budget, and so on

3.  Summary
        Summary of main points
        Date of next report
```

Figure 4-1. Sample outline of a progress report

about who, what, when, where, why, and so on. More important, trip reports should *interpret* the material they present. Any trip, no matter how routine, should be considered an opportunity to carry out an investigation, to be an active observer. The short example in Figure 4-2 shows how a sales representative turned a routine visit into an opportunity to research a new market.

INSPECTION REPORTS

Inspection reports should be organized to emphasize their primary purpose: to report on an inspection of specific features of some project or place. You might be asked to inspect a factory for fire safety or to evaluate the size and adequacy of a telephone network. While you should keep your eyes open for other important data, your inspection report should state your main purpose clearly and use that purpose as the organizing principle of the report. Figure 4-3 shows the introduction and summary of a report by a fire insurance underwriter. Notice how the primary purpose is kept clearly in view.

RECOMMENDATION REPORTS

Recommendation reports follow the basic three-part structure, with an introduction, a body, and a recapitulation. Because recommendations are generally requested by a supervisor, the outline of the body section includes several carefully considered features that help recommendations persuade their readers.

The introduction of a recommendation report should refer to the circumstances out of which the recommendation grows. If your employer asked for a recommendation report on a particular date, mention that fact. The introduction should also briefly describe the problem and state the recommended action.

```
                        MEMO


       TO:  Ollie Redburn
       FROM:  Alex Redding, Editing Division
       DATE:  November 15, 1985
       SUBJECT:  Marcom Computer/Editing Market


       On November 14 I drove over to Bellevue to
       discuss a manual writing project with Marty
       Williams of Marcom Computer.  I accepted the
       project for the company and left a contract.
       Marcom is involved in developing software
       manuals for trade book mass marketing.  I
       think there may be other small and medium size
       firms in the area that will be getting into
       this field more actively--in addition to the
       larger firms that are handling their own edit-
       ing, production, and sales.

       Marcom Computers is located at 27983 West 8th
       in Bellevue, 98111.  Telephone (206) 323-1897.
       Williams has a software guide under contract
       to Wilson Publishing with a submission date of
       January 1986.  He wants us to edit the text
       and review the organization between November
       20 and December 10.  He will deliver the text
       on the morning of the 20th with our signed
       contract.
```

Figure 4-2. Sample trip report

I suggest that we evaluate this project carefully with an eye to developing a long-term
relationship with Marcom; Williams was clearly
interested in developing his side of the business.

The text will arrive on the 20th with the
contract and a list of his questions and
guidelines. We should have the project finished by December 1st if possible; December 10
is the deadline.

<u>Sections</u> <u>of</u> <u>an</u> <u>Inspection</u> <u>Report</u>

On May 3, 1988 I made an initial visit to the
shop and storage yard of Agri-Spray, Inc. to
determine the level of risk and compliance
with applicable fire safety standards. We are
writing a new fire policy only. In this report, I will first present what I found and
then assess the risks.

[Body of report not reproduced]

At this time, Agri-Spray meets all fire and
safety standards, with the exception of a
paint cabinet that is 25 feet rather than 35
feet from the welding table in the engine
repair shop. I suggested that the cabinet be
moved or that the paint be stored at the far
end of the vehicle service shop where less
welding is done. I will check this on my next
scheduled visit in December.

Figure 4-3. Sample introduction and summary of an inspection report

The body of the report has four sections. (1) Describe the problem fully. (2) Describe the scope of your investigation and present the criteria that any solution to the problem should meet. (You describe the scope of your investigation so that the reader knows how your search for a solution has been limited by considerations of time, cost, or method.) (3) Present several reasonable alternative solutions, evaluating each according to the criteria you developed and presented earlier. Do not present your own recommendation until you have fairly evaluated all the alternatives. This order of presentation indicates that you are giving all solutions fair consideration, and it allows the solution you favor to have the last word. (4) Present your recommended solution and evaluate it carefully with respect to the criteria. The recapitulation or concluding section of the report restates the recommendation and presents its chief advantages and disadvantages. Remember that a recommendation is a suggestion, not an order. The final decision should be left to the reader who commissioned the report. You want to be persuasive, yet not overbearing.

Two common problems with recommendation reports concern *fairness* and *organization.* Readers easily see through a biased presentation. Present the advantages and disadvantages of all the alternative solutions fairly. An organizational problem arises because there are normally two attractive organizational strategies you can use in a recommendation report. If you are comparing three alternative solutions using four main criteria, you can discuss each solution in turn, treating the criteria as subheadings of each solution, or you can consider each major criterion in turn, treating the solutions as subheadings of criteria. Sometimes one pattern will be preferable, sometimes the other.

Because recommendations are meant to be persuasive, not merely factual presentations, you should consider how your audience is likely to react to the way you evaluate the alternative solutions. One important strategy has already been mentioned: Place your recommended solution last. If you anticipate that parts of your audience may disagree, you can please them by placing their favorite criteria first and by carefully pointing out the advantages of the solution they prefer. Keep in mind that while you are recommending one solution, you are rejecting several others. This bad news, like any bad news, needs to be buffered. The work sheet shown in Figure 4-4 will help you design an effective recommendation report.

FEASIBILITY STUDIES

Feasibility studies combine elements of recommendations and proposals. If you are asked to research the feasibility of some action or solution, your first question should be, "According to what criteria?" The question "Is nondestructive testing of our product feasible?" can have several answers. Nondestructive testing may be technically feasible, economically feasible, feasible in terms of existing staff and management expertise, or feasible in terms of available time and schedules. In every case, a feasibility study leads to the recommendation that something is or is not feasible according to the stated criteria. The outline in Figure 4-5 suggests a typical organization for a feasibility study.

CASES

PORTABLE COOLERS

In your two years as a mechanical engineer with Browning, Inc. you've been involved in several projects. The company specializes in ventilation, heating, and cooling systems for buildings and vehicles. Now, however, the company has embarked on a unique project under contract to Moorehead Chemical Corporation.

Moorehead has developed a special protective suit to be worn by personnel in the presence of hazardous chemicals. The suit is

1. My audience is

2. The criteria I have developed are

    a. Do these criteria need justification?

    b. The justification for my choice of criteria is

3. The alternatives I have studied are

4. The evaluation of alternatives in terms of my criteria is

5. My conclusions are

6. My recommendations are

Figure 4-4. Sample work sheet for a recommendation report

```
1.  Introduction and overview

2.  Body
    a.  Statement of problem
    b.  Statement of criteria for decision
    c.  Presentation of research on the
        question or questions (technical,
        management, cost, evaluation)
    d.  Evaluation of facts according to
        criteria
    e.  Conclusions and recommendations

3.  Recapitulation of main points

4.  Appendices
```

Figure 4-5. Sample organization for a feasibility study

specifically intended for use in the case of a chemical accident. To protect personnel from injury, the suit is completely impermeable and covers the body entirely. However, these very factors—impermeability and total coverage—create a major problem for those wearing the suit: Because the suit prevents air circulation, the wearer is unable to cool down by sweating. Heat stress is a real danger, and the suit cannot be worn for long periods.

Moorehead has accepted your company's proposed design for a portable liquid cooling system to be worn with the suit to prevent heat stress. Your group has been working on the system for the past eight months, both constructing a prototype and conducting extensive tests to prove its effectiveness. If Moorehead accepts your conclusions, your company will begin manufacturing the system within the next year.

During these eight months your group has submitted two progress reports to the managers at Moorehead. Now that construction and testing have been successfully completed, it is time to send Moorehead the final progress report prior to their first inspection of your finished product.

Your group has been assembling material for this report for several weeks. Because of your superior writing skills, you have been selected to write the final report. You have been given several minireports of varying degrees of formality by the other group members. The three minireports from which you need to get your material for your report to Moorehead follow.

System Design Report

The liquid cooling system has only two big components, one portable and one stationary. The stationary component is just a refrigerator for storing and freezing the ice packs. The portable system is our baby.

The portable system has three parts: a vest, an ice pack (heat sink), and a circulation system. You put

the vest on underneath the suit; then you put the pack
and the circulation system over the suit as a backpack.

The vest is net interwoven with plastic tubing. You
put it next to your skin and it cools down the torso,
shoulders, and upper arms (.52 square meters body area
out of 1.8 square meters total). Around one-third of
the body.

The ice pack has four kg. of ice and a copper tubing
system. Liquid flows through the tubing and you've got
the ice around it to keep it cool.

The circulation system is composed of a battery, a
pump, and a bypass control. The battery pack has an
open voltage of 10 volts; it's rechargeable. The pump
pushes the liquid through the vest and the ice pack
(the battery pack powers the pump).

The bypass control is a sort of thermostat or temperature control for the system. If you leave the valve
open, all the liquid flows through the ice pack to be
cooled, which is going to make the liquid drop to its
lowest temperature. But if you close the valve, no
liquid goes into the ice pack and the liquid warms up.
So you turn the valve to some point between open and
closed to get the temperature you want.

Liquid used is 50/50 mixture of propylene glycol and
water. This mix is used because propylene glycol has a
low freezing point. If you used just water, it could
freeze up in the tubes when it goes through the ice
pack.

Testing Report

All subjects voluntary and completely informed prior to experiments and prior to consent.

<u>Timing</u>: One experiment/day. Tests began at 8 A.M.

<u>Dress</u>: Male subjects wore shorts; females, shorts and T-shirts. All subjects placed rectal thermometers at depth of 10 cm. Subjects weighed, then instrumented for heart rate and four skin temperatures: chest, upper shoulder, thigh, calf.

Two test series (identical except for equipment worn by subject). Control-instrumented subject wearing only chemical defense ensemble. Cooling-instrumented subject wearing chemical defense ensemble plus cooling system.

<u>Procedure</u>: After assuming appropriate equipment, subject moved to thermal chamber and weighed. Thermal chamber heated to 38°C. If subject was wearing cooling system, it was turned on. Subject was moved to treadmill within chamber and required to walk in 15-minute intervals. Heart rate, skin and rectal temperature were recorded every 2 minutes. Between walks subject was allowed to rest for 3 minutes. Water offered. Amount of water consumed was measured and recorded to nearest 5 mL.

Metabolic rate was measured by measuring amount of oxygen expired during both work and rest periods.

Subjects were required to work at a rate of 400 kcal/hr (equivalent to 3.5-4 mph). Speed and slope of treadmill adjusted to reach required speed.

Runs continued until subject's rectal temperature reached 39°C <u>or</u> subject's heart rate reached 85% of age-predicted maximum <u>or</u> subject complained of extreme fatigue or heat stress <u>or</u> 3 hours were completed.

After the runs were completed, subject was weighed. Equipment and instrumentation were removed and subject weighed again.

Both series of tests completed September 18, 1985–January 25, 1986.

Test Results Report

The four reported skin temperatures were combined for an average temperature for both runs, control and cooling. For control runs the skin temperature began at 35.2°C, rose to 36.5°C after 10 minutes, to 37°C after 20 minutes, to 37.2°C after 30 minutes, to 37.4°C after 40 minutes, and to 37.9°C at 50 minutes when the experiment was terminated. For cooling runs the temperature began at 34°C, dropped to 33°C after 10 minutes, 33.1°C at 20, 32.9°C at 30, 32.7°C at 40, 32.5°C at 50, 32.8°C at 60, 33°C at 70, 32.8°C at 80, 33°C at 90, 33°C at 100, 32.8°C at 110, and 33.2°C at 120 when the measurements were terminated.

Rectal temperatures were also averaged for the control and cooling runs. Temperature during the control runs began at 37°C, rose to 37.1°C at 10 minutes, 37.5°C at 20, 37.8°C at 30, 38.2°C at 40, and 38.4°C at 50 minutes when the experiment was terminated. Temperature for the cooling run began at 37°C, rose to 37.1°C at 10 minutes, 37.3°C at 20, 37.3°C at 30, 37.8° C at 40, 38°C at 50 minutes where it remained (with a minor fluctuation to 38.1°C at 80 minutes) until the measurements were terminated at 120 minutes.

Sweat production and evaporation were recorded using the following formula:

Sweat evaporated = final fully clothed weight – initial fully clothed weight – amount of water consumed

```
     Sweat produced = final seminude weight -
         initial seminude weight
During control runs approximately 1.35 liters of fluid
were lost per hour.  Of this total, 33% evaporated.  To
replace total lost fluid, subject would have to drink 1
3/5 quarts of liquid/hour.  During cooling runs sub-
jects lost approximately 0.7 liter of fluid per hour.
Of this, 55% was evaporated.  To replace lost fluid
subject would need to drink 4/5 quart of liquid/hour.
     Time measurement:  all subjects wearing cooling
system lasted 3 hours.  Average for control runs was 45
minutes.
```

Report Conclusions

```
     System allowed subjects to work four times longer.
     System reduced initial skin temperature of subject
         by 0.5 °C.
     System did not reduce rectal temperature, but
         rectal temperature stabilized at 38°C within 120
         minutes (without system, temperature continued
         to rise).
     System reduced fluid loss by half and increased
         sweat evaporation.
```

Assignment. Write the final progress report for Moorehead Chemical Corporation using the information in these reports. Your audience is the company's chemical safety committee. They received two earlier progress reports, but the last one was sent three months ago and they will undoubtedly have forgotten some details. Remember to design appropriate graphics. (The graphics department will do the final versions.)

NEW GUARDRAILS

For the past two years your employer, R & C Construction, has been working on its biggest contract yet—installing new, safer guardrails on approximately 500 miles of state highways. The project began when the State Highway Commission decided that the guardrails currently in use represented a major hazard rather than protecting the state's drivers as they were intended to do.

Ideally, a highway guardrail should neither spear nor vault nor roll a vehicle in a head-on collision between car and guardrail. Moreover the end of the guardrail should not be a hazard in a collision for any size of vehicle, from semi-tractor trailers to subcompact automobiles. Before R & C began its replacement project, the majority of the state's guardrails had two types of end treatments: blunt-end and turned-down. On the blunt-end the end of the w-beam was simply cut off and either left straight or fitted with a curved end piece. This treatment resulted in a greatly increased hazard from spearing: in a head-on collision the end of the guardrail either punctured the passenger compartment or brought the vehicle to an abrupt stop which could result in further injury. As an alternative treatment, the state highway builders began turning down the ends of the guardrails, twisting the end 90° and sloping it into the ground. This modification took care of the spearing hazard, but it created a hazard of its own: testing showed that in a head-on collision the turned down guardrails could cause a vehicle to roll over or to vault over the guardrail as if it were a ramp. This vaulting, in turn, often sent the vehicle hurtling into precisely the hazard the guardrail was installed to protect against.

This was the point at which R & C entered the picture; the state requested proposals for replacement guardrails which would eliminate the hazards of the two current designs. After some preliminary research, R & C suggested three possibilities: a breakaway cable terminal, a modified turn-down, and the Sentre guardrail end. They proposed testing the three types, submitting the test results to the commission for approval, and then installing the best alternative on several highways chosen for upgrading. The results would then be evaluated by the State Highway Commission, which could decide whether to replace more of the old guardrails at a later date. R & C's proposal was accepted and the program was undertaken; it is now in its final stages, and it's up to you, as the project engineer, to write the final progress report to the state commission.

This progress report is to be a record of the entire project; it will be placed on file for public scrutiny. Consequently you must go back through your records and review all of your steps. It also means that you must explain some things you do not normally have to explain when you are writing for someone who has been involved in the project from the beginning. You decide it's time to review your notes, which follow.

Notes

Sentre system proposed on April 3, 1985. Proposal accepted on June 15, 1985. As of April 8, 1986, Sentre installed on 453 miles of highway; expected completion date: June 1986.

Breakaway cable testing (Jan-Feb 1985): Crash cushion-type end treatment. Absorbs vehicle impact in head-ons. Posts that support system break away on impact, guardrails crunch-up, absorbing impact. Cable provides tensile strength, holds guardrail end upright when guardrail hit from the side rather than head-on.

Results: Breakaway cable tested Jan-Feb 1985. Acceptable for vehicles 2,500-4,500 lb; no spearing, vaulting/rolling prevented. Spearing and rolling found with smaller cars. Not acceptable for 1,800 lb. Mod-

ification: slip-base steel posts rather than wood
(slip-base allowed posts to break away on impact). Results: severity of damage lessened, but damage still
occurred. Rated unacceptable.

Dates from 1970s. First developed with larger cars--
works well for them. Design didn´t take subcompacts
into consideration.

Turn-down modification: Advantages: took advantage of
existing hardware used with turn-down. Turn-down modification: remove bolts from first five posts supporting guardrail end. Guardrail held on with special
clips. Highway designers loved it. But they later
rejected it because it fell off the posts sometimes for
no apparent reason. Guardrail will drop on impact--
takes away vaulting/rolling effect. Tests: Sept-Oct
1984. Results: subcompacts still ramp on end and
roll; won´t work with 1,800 lb vehicle. Tests also
confirmed earlier reports that guardrails could fall
off posts without a collision. Another 70s innovation.

Sentre system: Collapsible, absorbs impact energy.
Steel slip-base posts, no snagging (wooden posts snag).
Developed in early 80s. Tests Nov-Dec 1984. Hit head-
on, telescopes to prevent spearing; built-in retaining
cable directs vehicle away from rigid end of guardrail.
No turned-down end, thus no ramping, vaulting, or rolling. Effective for all weight classes. Results:
acceptable.

Significant project dates
April 3, 1985: Sentre proposed
June 15, 1985: Sentre proposal accepted

```
July 2, 1984:  Original R & C proposal sent to State
    Highway Commission
August 20, 1984:  R & C proposal accepted
September 3, 1984:  Testing begins on turn-down modifi-
    cation
October 18, 1984:  Testing completed on turn-down mod-
    ification
January 4, 1985:  Testing begins on breakaway cable
February 13, 1985:  Testing completed on breakaway
    cable
October 29, 1984:  Testing begins on Sentre system
December 2, 1984:  Testing completed on Sentre system
July 10, 1985:  Installation project begins
June 1, 1986:  Expected date of project completion
```

Assignment 1. Write the progress report. Your audience is the State Highway Commission; it commissioned the project. Your purpose is to describe the progress of the project up to the present time and to predict future progress. Remember that your report will be open to the public.

Assignment 2. One of the members of the State Highway Commission has a special request for you after reading your progress report. He thinks the whole question of guardrail safety and the state's efforts to replace faulty guardrails would make a good article for the Highway Commission's semimonthly publication *Highways and Byways.* He asks you to write up a short article; your audience is the general reader, particularly travelers who drive around your state.

Assignment 3. A member of the State Highway Commission is interested in a concrete barrier (the New Jersey type) being used on the interstate highway system. He asks you to research that type for future reference. Research the design and performance of the New Jersey type concrete barrier and write a short informative report for the State Highway Commission.

COLLEGE FIRE ALARMS

You have just completed your first job as an electrical engineer for your new employer, Roytan Security Systems. Roytan manufactures a variety of alarm systems for both businesses and residences, and you've been working with Roytan's fire alarm systems. Your first assignment in your new position was particularly interesting since it came from your alma mater—Clemens University. For their new classroom building the Clemens administration wants the best alarm system they can afford and you think you've managed to fulfill their wishes; now it's time to let Clemens know what you've done.

The new building occupies 100,000 square feet including classrooms, offices, and a computer room. The Clemens administration would like a system that would include both detectors, warning devices, and some form of fire extinguishers; they would also like to spend no more than $70,000 if possible. As yet no final decisions have been made about the nature of the system to be installed. What Clemens wants is a study of the building with some recommendations as to the ideal system which they might use. The group which has commissioned your study is the Clemens Building and Grounds committee; the members include the Vice President for Campus Affairs, the chief engineer, and the architect for the new building, as well as professors from the biology, French, and marketing departments. Your job, then, has been to study the new building, consider existing fire alarm systems, and come up with a system which you think would do the job, preferably for less than $70,000. As you studied the problem, you were able to break the requirements down into four areas: detection,

notification, control, and coordination. Using these areas, you
have assembled the following data.

Detectors

Smoke detectors: Ionization type contains small amount
of radioactive material with electric current flowing
across it. When smoke reaches radioactive material:
1) electric current cut off; 2) alarm is triggered.

Photoelectric type has beam of light projected from
source to photoelectric sensor. Distance between
source and sensor ranges from less than 1 inch in
residential models to more than 15 feet in air conditioning ducts. Beam strikes sensor and produces
electric charge; if beam is blocked or diminished in
intensity, current produced by sensor also stopped or
diminished. This triggers alarm.

Heat detectors: Two types: fixed temperature and rate
of rise. Fixed temperature triggered when temperature
rises above some set level, usually 135 degrees F.
Most common type for office buildings. Rate of rise
triggered when temperature rises above rate of 4 degrees per second. Used where temperature of room is
normally high (machine rooms) or where there are com-
bustible fumes.

Pull stations: Alarms to be pulled manually. Usual
requirements: 9 stations/25,000 ft.

Limitations: Location of detectors regulated by
National Fire Protection Association (NFPA). Smoke
detectors no more than 60 feet apart. Should be placed
where there is constant air flow (e.g., hallways, air
conditioning ducts). Cannot be placed within 4 inches

of a corner; wall-mounted models cannot be more than 12 inches from ceiling.

Ionization detectors more reliable than photoelectric because photoelectric can be triggered by anything obscuring beam of light (e.g., surge of dust, object falling through beam, etc.). Ionization detectors average $85; photoelectric average $55.

Heat detectors have same location restrictions as smoke detectors. Heat detectors cost less than smoke detectors, can be placed in higher concentrations for improved detection. Average cost, $8 per unit.
N. B. Clemens building has no excessively hot rooms.

Notifiers
These simply notify personnel that: 1) there's a fire, and 2) they should leave building by fire exits. Include bells, horns, and red lights. Fire exits should be marked with red exit signs.

Controllers
After detection and warning, system should control fire. High expense; should be limited to rooms where there is expensive equipment and places where personnel may congregate during fire: lobbies, fire exits, etc.

Flow alarms: Basically sprinkler systems. Triggered by excessive heat. Suffocate fires with water. Used for wood and nonchemical fires; no good with electrical fires.

CO_2: Carbon-dioxide based extinguishers smother fires by preventing oxygen from getting to fire. Usually in form of fine white powder. Effective against electrical fires, some chemical fires.

Halon: Halon also prevents oxygen from reaching fire. Halon evaporates, which CO_2 does not. Leaves no residue. Good for rooms with expensive electronics, i.e., computers. Also good on some chemical fires.

Limitations: Flow alarm problems: everything in room is drenched; office machines and furniture ruined, professional cleaning required after fires. Cost high; average $1 per square foot of protection. Good for areas where people congregate. CO_2 leaves residue that can damage electronic circuits; must be cleaned professionally. CO_2 itself not expensive, but tanks and necessary tubing are ($8,000 to cover 3,000 square feet). Also dangerous to personnel; can result in asphyxiation for people in room when extinguishers go off (deaths have been recorded). Halon supposedly safe for personnel. Expensive: $8,000 for tanks and tubing plus $8 per pound for Halon (around 600 pounds needed for 3,000-square-foot computer room).

Coordinators

Need system to coordinate detectors, notifiers, and controllers. Best done by electronic monitors that sense when detector or extinguisher triggered and notify safety personnel of what's happened and where. Usually uses control panel divided into 12-16 squares with letters on each square. Each square refers to sector of building. Lights go on in sectors where detector has been triggered. Monitor also notifies fire department and triggers notifiers in building. Extinguisher monitor automatically triggers necessary extinguishers and notifiers.

<u>Limitations</u>: Extremely expensive (almost 50% of total cost). Must have separate alarms for detectors and extinguishers.

Cost Breakdown
  Ionization detector: $85
  Photoelectronic detector: $55
  Heat detector: $8
  Pull station: $35
  Bell/siren/horn/alarm light: $25 each
  Flow alarm: $1/square foot of coverage
  CO_2 system (for a single 3,000-square-foot room): $25/tank of CO_2 (6 needed); $8,000 for tanks and tubing
  Halon system (for a single 3,000-square-foot room): $8/lb of Halon (600 lb needed); $8,000 for tanks and tubing
  Monitors: $20,000/detector monitor; $15,000/ extinguisher monitor; $5,000/control panel

Physical Requirements
  Total area: 100,000 square feet
  Exits and lobbies: 5,000 square feet
  Computer room: 3,000 square feet

Assignment 1. Research the applicable regulations in your city for installation of fire detectors, notifiers, and controllers. Decide how many and what type of devices would be necessary for the new building at Clemens. Write a brief informational report on your findings.

Assignment 2. Write a recommendation report for the Clemens University Building and Grounds Committee describing the system you are recommending for the new classroom building. You need to explain why you arrived at the conclusions you did about each of the components of the system. Your audience also wants your cost estimates. Consider what graphics you might include to make your discussion easier to follow.

PROPANE-POWERED BUSES

During the past two years while you were working for your degree in mechanical engineering, you had a part-time job driving a school bus for the local school district. The work was enjoyable and the hours certainly fit well with your classes, but you never thought it would have any bearing on your engineering career. However, because of economic problems in your city, your school bus experience has given you a chance to exercise your engineering expertise.

Last year the school district went through one of its periodic budget crises; as a result all aspects of the current programs came under review, including school buses. The district has 1,065 buses that are driven more than 10,500,000 miles per year. Gasoline for the buses costs the district around $2,150,000 annually, and a 10 percent mileage increase is expected next year because of population growth. The district's budget analysts suggested that a substantial savings might result if the district could switch to a fuel other than gasoline. Their suggestion was diesel fuel, but you feel that diesel is not the best choice. Your junior project was a study of propane fuel systems for automobiles, and your research led you to believe that propane would be the best fuel for the district's buses.

You have discussed your opinion with both your major advisor and your boss at the school district. Your boss suggested you talk to the assistant superintendent of the district, Dr. Joanne Mirsch.

Assignment 1. Using the information provided in the research notes that follow, write a letter to Dr. Mirsch telling her of your interest and describing the advantages of using propane fuel for the school buses. This letter should be similar to a *proposal prospectus* (see page 158).

Research Notes

<u>Propane</u>: Internal combustion engines (usually modified gasoline engines). Storage tank for propane, hoses, and carburetor adaptor have to be added. Doesn't contain lead or sulphur, thus no gum or residue on engine parts. Longer engine and oil life. No condensation, exhaust temperature of 800°C. Decrease in mpg from gasoline (4 mpg average for propane versus 6 for gasoline and 10 for diesel). N.B. propane engines have safety valves inside; flow of propane stopped if there's a leak. Costs less than diesel or gasoline: average of $ .50 per gallon versus $1.15 per gallon for diesel and $ .99 per gallon for gasoline. Maintenance less also: $2,000 for 10 years versus $3,555 for gasoline and $6,000 for diesel. Reduced maintenance caused by lack of residue on engine parts, also oil changed less often due to less friction. (Gasoline requires oil change every 2,000 miles, propane 6,000 miles.)

<u>Diesel</u>: Diesel uses four-stroke engine. No spark plugs; compression and heat ignite fuel. Four strokes = inlet, compression, injection, exhaust. First, inlet valve opens, lets air in cylinder. Then air compressed and becomes hot. Fuel sprayed ("injected") into hot air--ignites and burns. Creates high pressure that forces piston down. Then burned gases forced out ("exhausted"). Diesel compression ratio higher than gasoline engines; higher ratio allows more torque at low engine rpm's. Diesel puts more power

into drive train of vehicle than gasoline or propane. Diesel cylinder walls heavier to stand pressure of ignition, more metal used. This reduces power-to-weight ratio. Diesel exhaust smellier and smokier than gasoline or propane. Diesel safer than propane or gas because it requires higher temperature to ignite.

Gasoline: Also four-stroke, but uses spark plugs to ignite. "Internal Combustion" engine. Gas-air mixture enters cylinder in inlet stroke; compressed and ignited by spark plug in compression stroke. Pressure pushes piston down in power stroke; burned gases forced out in exhaust stroke. Fuel may ignite before spark plug sparks--"preignition." Requires engines with small compression ratio (no more than 11 to 1). Reduces power of engine applied to drive train. High exhaust temperature--900°C--also gas can condense. Gasoline components--lead, sulphur, etc.--form residue on engine parts when heated, interferes with function of engine.

Weather effects: Heat: biggest problem in this location (10 school days over 100°F in August/September of last year). All three fuels OK in heat as long as cooling system kept supplied with coolant. Rain: Internal combustion engines can stall if distributor cap gets wet (spark can't ignite fuel). Diesels not affected by water (no distributor). Internal combustion engines OK if you watch it when you drive through puddles. Cold: Diesels have a hard time starting; they require heat to ignite fuel. Gasoline engines can have problems starting if fuel has condensed. Propane doesn't need heat to start, doesn't condense. Easier to start in winter.

Costs: Diesel bus costs $40,000; gasoline, $35,000; propane, $36,000. Propane is gasoline engine bus with

```
propane system added.  Propane system costs $1,000.
Current resale after 10 years:  diesel, $49,800; gaso-
line, $45,555; propane, $41,000.  Propane system is
removed from bus before resale; system can be installed
on a replacement bus.  Total fuel cost over 10-year
period (estimated 8,000 miles per year):  diesel,
$13,800; gasoline, $19,000; propane, $15,000.
```

Dr. Mirsch was interested in the information in your letter; she has asked you to write her a detailed report on the matter. After conferring with your advisor, you've decided to make this your semester project.

Assignment 2. Write a recommendation report on the conversion of the school system buses from gasoline to propane. Your audience is, again, the assistant superintendent of schools, Dr. Joanne Mirsch. Consider what information would be most interesting to her; also consider what graphics would best help convey the information in the report.

GEOTEXTILES FOR HIGHWAYS

You have been hired as an assistant director of your state's highway construction department. Your job is to oversee the department's construction design division.

The design division handles more than 5,000 miles of both paved and unpaved roads per year throughout the state. The division staff seems to be very competent; your only complaint thus far is that the members seem somewhat conservative. There is some resistance to new ideas and new developments.

The latest case in point is the matter of geotextiles. While you were doing your graduate work, you read several studies about the use of these materials in highway construction; they seem to be very effective in certain situations and can reduce the costs of future maintenance. Yet no one in the design division seems familiar with them.

You decide that the solution is to write a short report outlining the characteristics of geotextiles and their applicability to future projects for the division. First you do some research and find the following data.

> **Geotextile Data**
>
> <u>Geotextiles</u>: Membranes used to stabilize soils. Types include polymers, membranes, and synthetics. <u>Polymers</u> include polyamids, polypropylene, polyester, and polyethylene. <u>Membranes</u> include woven, knitted, and nonwoven heatbonded fabrics. Also nonwoven needlepunched fabrics, meshes, webbings, and mats. <u>Synthetics</u> made from wood pulp (rayon and acetate), silica, (fiberglass) and petroleum (synthetic rubber and polyvinyl chloride). Synthetics and polymers resist rot and chemicals, but exposure to sunlight can damage some fabrics (e.g., polypropylene). Ultraviolet stabilized polypropylene available. Fabrics can be custom designed and treated.
>
> <u>Four main functions</u>: Separation, filtration, drainage, and reinforcement. Fabric can be used to separate materials (<u>separation</u>), can allow water but not soil to pass through (<u>filtration</u>), can allow water to pass through without erosion (<u>drainage</u>), and can increase tensile strength (<u>reinforcement</u>). Main function of fabric is to prevent natural forces from moving exposed surface soil. Water permeability must be matched to soil being protected. If soil and geotextiles have equivalent permeability, geotextiles can have up to one thousand times greater flow capacity than soil.
>
> In unpaved road construction, geotextile layer can be placed between aggregate and subgrade soil levels.

Separation and reinforcement qualities combine to sta-
bilize surface; filtration and drainage qualities combine to allow water to pass through surface without
causing erosion or mixing layers. Performance also
improved. Thickness of aggregate layer can be reduced;
more cost-efficient road.

Also good for asphalt. Improves resistance of pavement
to reflection cracking; creates moisture barrier that
protects underlying pavement from further degradation.
Fabric's structure determines threshold asphalt level
needed to form moisture barrier (dense, thin structures
require less asphalt than thick porous ones). Good
reinforcement performance, but tack layer necessary.
(Tack coat = tarlike substance sprayed on asphalt to
glue layers together.)

Costs vary depending on quality, weight, fiber, type of
manufacture. Nonwoven polypropylene ranges from $ .75
to $2.00 per square yard. Woven costs the same. Geo-
membranes go from $1.50 to $5.00. Lighter weight textiles cheaper; heavier weights more expensive.

Next you need to compose the report, but you can see some
potential problems. The design department is made up of eighteen
people; three of them have undergraduate degrees in engineering;
the others are all technicians with high school and junior college
degrees in various disciplines. However, all of those on staff are
more experienced than you are; the average number of years on
the job is seven. You don't want to alienate these people by giving
them the impression that you undervalue their experience. At the
same time, you'd like them to be more aware of new developments in the field. Somehow you need to get the message across in
a way that's nonthreatening but definite enough so that the report
is not ignored.

Assignment 1. Write an informative memo to acquaint the members of your staff with the uses of geotextiles.

Assignment 2. Now take this information one step further. Rather than an informative memo, write a short recommendation report on geotextiles for the members of the highway design division. Your purpose is to recommend that geotextiles be incorporated into designs when they're appropriate and to provide enough information so that the staff will know enough about them to know when it's appropriate to consider using them. Obviously they'll have to get more information when they actually design the roads.

NONDESTRUCTIVE TESTING

You have recently been employed by an electronics firm that produces plastic daisy wheels, metal casings for computer printer print-heads, and circuit boards. In an increasingly competitive market, the company wants to reduce the number of defective parts it sends out, and it hopes that it can also increase the average life of its products by identifying parts that contain small flaws—flaws that will not result in immediate failure but will reduce the life of the part.

Next month, as part of the annual budget review, the company's Capital Expenditures Committee will meet to discuss new equipment purchases. Your supervisor has asked you to prepare a preliminary report—including recommendations—that reviews the possibility of applying nondestructive testing methods to the company's plastic, metal casting, and composite products. Your supervisor believes that some nondestructive testing technology can be applied in the company, but she is not sure which technology is most cost effective or whether any of these technologies can be routinely applied to all the products that come off your assembly lines.

Nondestructive testing includes any method for testing a product that does not destroy the item being tested. For example, nondestructive methods for testing plastic pipes might include x rays or ultrasonic tests. One advantage of nondestructive testing is that every piece that comes off the production line can be tested. By contrast, with *destructive* testing methods a company randomly selects a statistically significant sample of the product to generalize about overall quality. This method offers no assurance about the quality of any specific item.

Your boss places one important limitation on your research. She does not want to have salespeople hanging around the front office trying to promote a particular system before the Capital Expenditures Committee has had a chance to inform itself about the possible systems, their costs, and their effectiveness. Consequently, you are to do all your work by consulting journals and other printed publications. No letters, no phone calls. Above all, no discussions with competitors or suppliers because your company does not want to spread the word that it is concerned about product quality and reliability.

As far as you can tell, the members of the committee are generally neutral on the subject of nondestructive testing, with the exception of one member who feels that it is a sufficient test of quality to select random parts from the assembly line and submit them to destructive tests by running them until they fail. In addition, the committee has several other major expenditures to consider before the proposed budget is sent to an executive committee for review. It is unlikely, therefore, that the Capital Expenditures Committee will recommend installation of a system next year. Instead, it will probably recommend further research on a particular system. Your report should be about fifteen highly focused pages.

Your boss has collected a short bibliography of materials on the subject:

Nondestructive Testing. Technology Utilization Office, NASA, Washington, D.C., Publication Sp5113, 1973. National Technical Information Service No. 3300-00471.

Proceedings of the 14th Symposium on Non Destructive Evaluation. Southwest Research Institute and American Society for Nondestructive Testing, San Antonio, Texas, April 19–21, 1983.

Real Time Radiologic Imaging: Medical and Industrial Applications. American Society for Testing and Materials, Philadelphia, Publication STP-716.

Digital Nondestructive Evaluation of Composite Materials. Robert Blake. Center for Composite Materials, University of Delaware, May 1982.

Nondestructive Testing of Fiber Reinforced Composites. George Matzkanin. Southwest Research Institute, Office of Nondestructive Evaluation, National Bureau of Standards, 1982.

Assignment 1. Research the subject of nondestructive testing for your boss, using these sources and others. Write an annotated bibliography of sources on the subject—a list of sources with a brief summary and evaluation of each.

Assignment 2. Write the report on nondestructive testing for the Capital Expenditures Committee, keeping in mind the constraints mentioned in the case.

THE INSPECTION REPORT

With the help of your instructor, identify a subject for an inspection report. Consider the following possibilities:

- A damaged piece of lab equipment (purpose: identify needed repairs).
- A personal possession in need of repair (purpose: identify needed repairs).
- A building site near campus (purpose: identify work required for clearing and possible uses of the property).
- A feature of your campus or work site in need of maintenance (examples: landscaping, coffee rooms, typing or lab facilities).

Assignment. Write an inspection report on your chosen subject. State your main purpose clearly and organize your report around that purpose. Remember, inspection reports should concentrate on the specific features of some project or place.

THIN-PANEL DISPLAYS

For the past three years you have been the head of a design group at Climax Electronics, an established producer of electronic components. During the last six months your group has been working almost exclusively on thin-panel display devices—small screens that can be used for applications ranging from portable computers to digital watches. Climax intends to market both instruments with thin-panel displays and the displays themselves as components for other manufacturers. You've been concentrating on three types of displays: liquid crystal, AC-gas plasma, and electroluminescent.

Last month, however, Climax was sold to another corporation —Wholesome Products. Wholesome began several years ago as a canned food processor, but it has since grown into a large diversified corporation with holdings in a variety of fields.

Problems began for your group when the Wholesome management decided to review all current projects at Climax. You were informed that your research and development budget was being reduced as a cost cutting measure. The Wholesome management feels that work on three thin-panel devices is unnecessary, and it wants you to suggest alternatives for the current program. It has ordered you to eliminate one of the devices from the project, which means a careful review of the work you have done on the three displays.

You've asked the three members of your group—Maureen Klausewitz, Larry Greener, and Chris Montrose—to evaluate the project thus far with a view to deciding which display device deserves more emphasis and which device(s) might be eliminated from the project. They are to submit memos to you with their conclusions. You also ask Larry Greener to give you a quick review of the working principles of the three devices for reference purposes. Here is Larry's memo to you, Joanne Huerta.

```
TO:  J. Huerta
FROM:  L. Greener
SUBJECT:  How the Display Devices Work


In reviewing the working principles, I'll start with
the liquid crystal displays (LCDs), although you prob-
ably remember most of this.  These are the most common
displays, of course, everything from digital watches to
appliance control panels.

The LCD is basically two polarizers surrounding some
liquid crystal material (LCM).  This LCM is made up of
rod-shaped molecules that are normally in some kind of
spiral configuration.  When you apply an electrical
field to the LCM, though, the molecules straighten up
and become parallel.  The two polarizers are set up so
that their transmission axes are perpendicular (the
axis of the bottom is perpendicular to the axis of the
```

top) and there's a reflector below the bottom polarizer.

When the LCD is off (i.e., there's no voltage to the
pixel--the picture element or cell of the display
device), the spiral configuration of the molecules
scatters polarized light from the top polarizer; this
orients the light so it can go through the bottom
polarizer. Then the light hits the reflector and goes
back through the bottom polarizer the same way. This
makes a "lighted" panel. When you turn the LCD on, you
put some voltage on the pixel and that creates an
electric field. That makes the LCM molecules
straighten out and become parallel, as I said. When
that happens the light is no longer scattered and all
the polarized light can be absorbed by the bottom
polarizer. No light goes to the reflector and you get
a "darkened" pixel. The dark pixels are what show
information on the LCD screen.

The AC-gas plasma screens are really glass envelopes
full of some low-pressure gas like neon. <u>Plasma</u> refers
to the <u>glowing</u> neon gas that gives off light when the
screen is turned on. You've got conductors around the
gas and there's a spacer to keep the glass walls apart
because the gas is under such low pressure that the
walls could collapse from the atmospheric pressure. If
you apply a high enough voltage (threshold voltage) to
the gas, it excites the electrons which release energy.
This causes a bright, orange-red light. The screen has
a matrix of electrical conductors in rows (horizontal)
and columns (vertical), and it has a gas-plasma cell at
each place where the rows and columns intersect. When
the display is off (normal state), you've still got

electricity applied to all the conductors (row and column), but the combined voltage at the intersections is less than at the threshold so the cells stay dark. To light them up, you increase the voltage until you get threshold voltage at the intersections, which makes the cells light up. If you want to turn a cell off, just lower the voltage on the correct row and column. The lighted cells are what show information on the screen.

The electroluminescent display (ELD) is similar to the AC-gas plasma in terms of how it works. The screen is also a matrix with the row and column conductors with an ELD at each intersection. The ELD is made up of a phosphor between two layers of glass. The glass has conductors etched into it. The phosphor is usually zinc sulfide (ZnS) doped with manganese (Mn). If you give the phosphor a threshold voltage, the Mn electrons get excited and give off light that's yellow-amber.

The memos from the group members about the three thin-panel display devices follow.

TO: J. Huerta
FROM: M. Klausewitz
SUBJECT: LCD versus Gas Plasma

Let's face it, LCDs are used very widely because of their advantages—they're easy to produce, they're cheap, and they don't require much power. The fact that they're easy to produce is reflected in the quality control statistics: Only one out of ten LCDs produced is defective. That's one of the reasons you see so many LCDs in watches and calculators and portable computers, for example. And because they're easy to produce, they're cheap—around $400 for a 16-line

display. Because LCDs are light absorbers and get
their light from other sources, they don't have to
consume power to make light. A 16-line display only
uses about 1/8 watt of electricity. That's what makes
LCDs so good for portable computers--you can run a
portable on a battery pack for about 10 hours without
recharging.

Compare this to the AC-gas plasma display and you can
see why the LCD is more common. Gas plasma is a light
emitter so it needs a lot of power to break down the
neon gas. A 25 x 80 line screen would take 30 watts,
which means the gas plasma won't work for portable
computers. The gas-plasma screens are hard to produce
too--the glass has got to be thick to stand up to the
pressure of the atmosphere against the low pressure of
the gas. You have to use spacers to keep the glass
from bending, which interferes with light emission and
reflects light from some of the pixels. Some of the
"off" pixels can look "on" because of light reflected
from the spacers. Because the gas-plasma screens are
hard to produce, they're also really expensive--about
$1,600 for a 25-line display. So far they're only
being used for some large screen displays like military
battlefield displays.

The LCD seems to be the display of choice and since our
resources will be limited in the future, I'd say we
should concentrate our research and development there.

TO: J. Huerta
FROM: C. Montrose
SUBJECT: LCD Problems

LCDs are common, but they're not faultless and we
should look at them very carefully. Basically the

problems with LCDs all come out of their display
abilities. First of all, because LCDs are light
absorbers, they don't give off any light; that means
you can't use them in low-light situations unless you
use a backlight, which is costly for power. A backlight for a 16-line screen would consume 1 to 2 watts
of electricity. There's also a problem with the contrast ratio (the ratio between the maximum and minimum
brightness levels) for the LCD: It's 5:1 which is very
low. Compare that to the 25:1 contrast ratio for a
cathode ray tube (CRT) display such as a standard TV.
The bad contrast ratio means that it's hard to look at
an LCD for any length of time. There are also problems
with resolution--the ability to differentiate between
similar images. The more lines of pixels per inch with
a display screen, the better the resolution. The resolution of the LCD is around 30 lines per inch, which
doesn't give a sharp picture. Finally, the viewing
angle of the LCDs is restricted to a 25-30° area. You
can't really see the picture well outside this viewing
angle.

Beyond these display problems, another big problem with
the LCD is that it isn't particularly sturdy. You can
only use an LCD in temperatures that range from 0 to
50°C (32 to 122°F). Outside that range the liquid
crystals transform and won't work. The screens themselves are pressure sensitive; if you press on the
screen (not the glass cover, but the screen underneath), the picture will fade out.

The last point is that the LCD screen is slow to
respond (1/4 to 1/2 second) and the picture has to be
renewed constantly even if it doesn't change. That

means LCDs just don't work for large screens--they
flicker.

I'd have to say LCDs have come and gone; there's got to
be something better.

TO: J. Huerta
FROM: M. Klausewitz
SUBJECT: Gas-Plasma Problems

Looking over our data again, I've discovered a few more
problems with the gas-plasma displays. The extra-thick
glass on the display also leads to weight problems: A
25-line screen weighs about 3 pounds. It's also about
3 inches thick. That doesn't seem like much compared
to a cathode-ray tube device, but compared to other
thin-panel displays like the LCD it's significant. The
LCD is much smaller and lighter.

TO: J. Huerta
FROM: C. Montrose
SUBJECT: AC-Gas Plasma Displays

I admit that the gas-plasma displays have some problems, but we shouldn't overlook their very real advantages. The gas-plasma display characteristics are very
good, particularly compared to the LCD. The gas-plasma
display is just as bright as the standard CRT--it can
give off light ranging from 40 foot-lumens to 200 foot-
lumens. You've also got a contrast ratio of 20:1;
that's very close to the CRT's 25:1. Then you've got
terrific resolution: 70 lines of pixels per inch.

The gas-plasma display isn't affected by temperature changes and it doesn't need to be renewed, which means it's ideal for large displays.

There's a real possibility here for a high quality device.

TO: J. Huerta
FROM: L. Greener
SUBJECT: Electroluminescent Displays

I've been looking over the recent memos and it seems to me we're stuck in a morass. Let me offer a way out: the electroluminescent display. It's been around since the late 1950s, but there were problems with screen brightness and lifetime. These problems have been ironed out and I think it's a real contender.

First of all, the ELD is about 75% as bright as a CRT; the light ranges from 20 to 30 foot-lumens. The contrast ratio of 20:1 is also good. It's got good resolution at 66 lines of pixels per inch. Finally, the most exciting thing about ELDs is the possibility of a multicolored screen. This would be produced by different phosphorous materials that give off different colors.

These screens are rugged, with solid cells, and they can run on a battery pack for 2 or 3 hours without recharging. Finally, the screen image on an ELD is sharper than that on a CRT.

Let's consider ELDs carefully; they may present a solution to our problems.

TO: J. Huerta
FROM: M. Klausewitz
SUBJECT: Problems with Electroluminescent Displays

I agree that ELDs do offer some exciting possibilities,
but we shouldn't let those blind us to their problems.

First of all, there are the same old problems with
power. Like the gas-plasma display, the ELD is a light
emitter, which means it needs power: 13 watts for a 25
x 80 line screen. It's also a real headache to produce. The phosphor has to be applied almost perfectly,
and it's very susceptible to defects like dust. Dust
makes the display useless. Quality currently is very
low--only 1 out of 4 ELDs produced can be used. This
causes high prices: $700 for a 25-line display.

Right now it doesn't look good to me.

TO: J. Huerta
FROM: C. Montrose
SUBJECT: Electroluminescent Displays

Larry has pointed out some terrific things about ELDs,
but they need a lot more work. Right now ELDs have a
limited lifetime and that lifetime decreases as you
increase the brightness of the screen. The average
lifetime is 40,000 hours. The multicolored screen is
still only a possibility; the phosphors need to be
about 10 times brighter than they are now for a multicolored screen to be feasible.

At least we know that gas-plasma displays work since
they're currently in use.

Assignment 1. Prepare a short (ten minutes) oral presentation, audio and visuals, on the three thin-panel displays you're currently working on to be delivered to the new management of your division. Consider what aspects of the displays are of most interest to this audience; also consider what visuals you need to help explain this material to an audience with little background on the subject.

Assignment 2. Based on the information in the memos, write a recommendation report for the management at Wholesome Products. You can decide to concentrate on only one of the display devices, or you can divert most of your budget to one while continuing work on another. However, at least one of the devices must be eliminated from the project. Consider how much the managers at Wholesome need to know about the devices to understand your recommendation. Also consider what graphics you might use to help clarify your report.

Assignment 3. Research the current status of the three thin-panel display devices. Have there been any new developments in the ongoing research? Write a short informational report on your findings.

NUTS AND BOLTS

Last week, you started your summer internship with Pan Engineering, a manufacturer of gardening tools and groundskeeping equipment. Pan's products range from home lawnmowers to machines used to prepare the surfaces of stadiums and ice rinks. Your supervisor, who is in charge of assembly operations, took you on a tour of the parts department, where you saw hundreds of bins of various screws, bolts, snaps, and other connectors. He has asked you to prepare a short memo that is due tomorrow on what he can do about this "situation." The supervisor seemed annoyed by the large number of fasteners and the somewhat confusing variety of basic types. His parting comment was: "Look at the duplication! The waste! Do you know how much this stuff costs?" After checking some texts and the catalogs of a few suppliers, you come up with the following information.

```
The Data
1.  Fasteners can be rendered corrosion resistant and
temperature resistant by coating and electroplating
processes.
2.  A bolt is an externally threaded, cylindrical fas-
tener that has a head that can be turned by some vari-
ety of tool.
3.  For maximum strength and efficiency, bolts need to
be tightened into what is called the yield zone.
4.  Threaded nuts and bolts were first produced in the
fifteenth century.  Bolts are fasteners, as are nails,
screws, rivets, spring clips, and other devices.
5.  There are many standards for bolts developed by
various national and international agencies, including
the ISO, SAE, ANSI, and the American Society for Test-
ing Materials.
6.  Aluminum plating provides temperature resistance.
7.  Zinc and cadmium plating make bolts corrosion
resistant.
8.  Standards have been developed to assure that bolts
have strengths and endurance adequate for their in-
tended uses.
9.  Bolt heads can be flat or convex.
10.  There are at least 8 types of head designs, in-
cluding the round, flat, oval, pan, fillister, truss,
hex, and washer styles.
11.  A nut is a block of metal with a hole in its
center that is internally threaded.
```

12. Nuts come in a variety of shapes, including hex, square, lock, cap, wing, and castle designs.
13. If the nut comes off, the fastener will often fail.
14. Nuts are designed and treated in many ways to prevent loosening.
15. Leonardo da Vinci drew designs for bolt-making machines.
16. If a product design calls for many different types of fasteners, costs can rise quickly and assembly can become inefficient.
17. Bolts have many drive systems, from the common slotted-head to the Phillips drive, the countersunk hex drive, the hex head, the square head, and the security head, which can only be driven in one direction.
18. Cadmium is a dangerous metal and has been associated with environmental hazards.
19. The security head bolt is generally used to prevent vandalism.
20. One authority says that any operation that spends more than one million dollars a year for fasteners can save 15% by keeping a watch on the variety of fasteners used and the efficiency with which they are used.
21. Zinc and cadmium plating make metal brittle.
22. If a smaller bolt is properly tightened, it can often be as effective as a larger bolt in many applications. These substitutions can save up to 10 cents per fastener.
23. The yield zone is far from the breaking point. It is a stage at which a bolt begins to stretch.
24. Special wrenches are available that allow close monitoring of the torque to which bolts are tightened.
25. Slotted nuts and castle nuts have openings for cotter pins, which are used to keep the nut in place.

26. Washers come in many shapes and sizes and include lock washers that keep nuts from loosening.

27. Two standard books on this subject are the Guide for Criteria for Bolted and Riveted Joints by J. W. Fisher and J. H. A. Struik and the Standard Handbook of Fastening and Joining, edited by Robert O. Parmeley.

Assignment 1. Write the memo the supervisor asked for, but first consider what kind of audience he is and what purposes he might have in asking for the memo. Some of the information in your list of facts is incomplete. Your instructor might ask you to perform more research in one or more of these areas, but what area might you offer to research further? Again, consider audience and purpose. Organize your memo as effectively as you can.

Assignment 2. Your boss is interested in your recommendations. Research and prepare a ten-page recommendation report on your most important finding in Assignment 1.

Assignment 3. Your boss is pleased with your recommendations, but he would like you to expand your research to the latest types of fasteners. "Are there cheaper, tamper-proof fasteners that are just as strong as a good small bolt?" he asks.

THE LANDFILL GAS STUDY

For the past ten years, while working for the city of Meridian, you've watched your town grow from a middle sized city surrounded by agriculture to one of the state's major manufacturing centers. Although the growth has brought many benefits to Meridian, it has also produced the inevitable problems. One is that the expense of city services has risen sharply and seems certain to continue rising. Two of the major demands on the city are your concern: power generation and waste disposal. Not surprisingly, you've been considering ways to combine the two.

Most of the city's waste is disposed of in sanitary landfills: large pits lined with impermeable membranes where the waste is dumped, compacted, and covered each day with a layer of earth. The landfill method is obviously superior to the old-style open burning that created air pollution and rodent hazards, but it has problems of its own.

The principal hazard with the sanitary landfills is the production of methane gas. The solid waste in the fill begins to decompose as soon as it is dumped; this decomposition produces carbon dioxide, hydrogen sulfide, a small amount of ammonia, and methane. After six months the waste produces 50 to 70 percent methane gas, the rest being carbon dioxide. The methane will continue to be produced for several years depending on the amount of oxygen present (oxygen inhibits methane production), the amount of moisture in the waste (60 to 80 percent moisture produces maximum methane), the pH in the environment (methane requires a neutral environment; most landfills are fairly acidic at first), and the chemical composition of the waste (biodegradable materials such as garden and kitchen waste, textiles, and wood stimulate a high rate of methane production).

The methane will migrate through the landfill as long as there are pores big enough for it to enter. If the surface of the fill has been paved or is frozen or saturated with rain, the methane will move laterally through the adjoining soil. When the methane finally escapes from the landfill and mixes with 5 to 15 percent air, it becomes explosive.

You think there is a solution to the problem of methane production in the form of landfill gas recovery. The idea is that the methane is recovered from the landfill and used to produce power for the surrounding area. At the moment your interest in landfill gas centers on one local landfill, the Dover Gardens site. Dover Gardens has around 400 acres with 250 acres used as a landfill approximately 30-70 feet deep. It is one mile outside the city limits, surrounded by open country, farms, a few industries, and a small residential area. Originally Dover Gardens was owned and operated by a private company—C D Industries—that sold the landfill to the city three years ago. It is now Meridian's principal landfill site. Your research about Dover Gardens and landfill gas production has uncovered the following facts.

Methane wells drilled in landfill to between 50% and 90% of landfill depth, connected to a gathering system. (Methane extracted through a perforated pipe inserted into well and connected to vacuum pump).

Well filled with gravel and sealed at surface. Each well connected to control valve and collection pipe that runs to gas-processing plant nearby.

Unpurified methane can be used as is for generating electricity--fueling turbines or internal combustion engines or steam boilers to run generators. Use of unpurified gas requires construction of a power gener-ating plant at landfill; power generated can be sold to local utility company.

Use of landfill gas recovery not always economical; must be user nearby who can make use of all gas, 24 hours a day.

For other uses methane must be purified, removing car-bon dioxide and other gases to upgrade it to high-grade methane (950-1000 Btu) that can be injected into exist-ing natural gas pipelines and used in same applica-tions.

Use of existing pipelines can reduce cost of landfill gas; meters can be used to measure amount of gas delivered and to insure that methane is consistently high grade.

Current trash amount shipped to Dover Gardens site: 1800-2200 tons/day. Estimated annual tonnage: 1986, 1.7 million; 1987, 2.5 million; 1989, 4.7 million; 1991, 6.2 million.

To find potential total rate of landfill gas production: Multiply tons of solid waste by high and low estimates of standard cubic feet (scf) of gas/ton of waste (standardized to low of 80 and high of 280 scf/ton).

1989: Dover will have 4.7 million tons in place. Estimated production for year:
  Low: (4.7 million tons) (80 scf/ton) = 376,000,000 scf/year
  High: (4.7 million tons) (280 scf/ton) = 1,316,000,000 scf/year

Daily rate = annual rate/365 days:
  Low: 376,000,000 scf per year/365 days = 1,030,137 scf/day
  High: 1,316,000,000 scf per year/365 days = 3,605,480 scf/day
(Only approximate; only on-site extraction tests can be accurate)

Costs to finance and operate electrical generation facility (using unpurified methane) vary from $2 million to $4.2 million; average is $3.1 million.

Costs for high Btu production vary from $1.6 million to $6.7 million; average is $3.85 million.

Cost of electricity from utility is around 6¢/kwh for industrial users.

Gas price around $4.38 to $4.66/million Btu for industrial users; average is $4.40.

Gross revenues for electricity generation (lowest quality methane) based on gas production rate, Btu content, price received for energy, and energy efficiency of operation.

Multiply high and low gas production rates by Btu content of gas (usually 500 Btu/scf), by the price to be received, and by the efficiency factor (amount of energy delivered to user divided by total amount of energy collected):
  (gas production rate)(500 Btu/scf X efficiency factor
   X gas price)

Low: (376,000,000 scf/year)(500 Btu/scf X 0.90 X 4.40/million)
   = (376,000,000 scf/year)($0.00198 per scf)
   = $744,480/year

High: (1,316,000,000 scf/year)(500 Btu/scf X 0.90 X 4.40/million)
   = (1,316,000,000 scf/year)($0.00198 per scf)
   = $2,613,600/year

For high Btu natural gas production, calculation the same except that efficiency factor is lower (greater energy required in processing):

(gas production rate)(500 Btu/scf X efficiency factor
X gas price)

Low: (376,000,000 scf/year)(500 Btu/scf X 0.75 X
4.52/million)
  = (376,000,000 scf/year)($0.001695)
  = $637,320/year

High: (1,316,000,000 scf/year)(500 Btu/scf X .075 X
4.52/million)
  = (1,316,000,000 scf/year)($0.001695)
  = $2,237,400/year

Gas production can last 10 to 15 years. Multiply gross
revenue for high and low by 10 minus cost for each
method.

Unpurified methane electricity generation:
  (Gross revenue X 10) - cost of method = net revenue
  Low: ($744,480 X 10) - ($3.1 X 10^6) = $4,344,800
  High: ($2,613,600 X 10) - ($3.1 X 10^6) = $23,036,000

High Btu conversion:
  Low: ($637,320 X 10) - ($3.85 X 10^6) = $2,523,200
  High: ($2,237,400 X 10) - ($3.85 X 10^6) =
  $18,524,000

Meridian Light and Power can buy either high Btu methane or generated electricity.

Assignment 1. Write a feasibility study for the Dover Gardens
landfill gas production project. Your study will be directed to the
city manager, who is in charge of decisions about the economic
future of the city. Consider which of your facts would be most
interesting and relevant to him. (Incidentally, he has very little

background in waste disposal.) Your purpose is to discuss the feasibility of producing landfill gas at the Dover Gardens site and to suggest alternatives for the production if you decide it is feasible.

Assignment 2. The mayor is interested in the costs and the payback aspect of the plan. He wants a short memo from you about how much the city can expect to spend and get back in return. The mayor, although extremely popular, is notoriously unskilled in mathematics. You need to explain costs and benefits of the landfill gas project to him, but you know you'll have to do it pretty simply.

Assignment 3. The city council is also interested in the plan, but the members don't like reading reports. Write a script (audio and visuals) for an oral presentation of the plan to the council. They're busy and will give you only fifteen minutes. Decide what would interest them the most and what material is vital to get your point across.

THE TRIP REPORT

For this case you must choose an event in which you are interested and which has some bearing on your career plans and write a trip report based upon it. Possibilities might include:

- A local or regional meeting of a professional association
- A professional conference
- A trade show
- A lecture or presentation
- A seminar at another institution

Assignment 1. After attending your chosen event, write a trip report on it. Be sure to include all of the relevant facts about the purpose of the event, what took place, who was present, and when and where it was held. Be specific about the part of the event that is relevant to your own career plans. Interpret the event for your instructor in light of those plans.

Assignment 2. Describe the event in a ten-minute oral presentation to your instructor and your classmates.

THE HEAT PUMP

During your three years at college you have never really had a chance to use your new skills on anything except class assignments. You've assumed that real life applications would have to wait for graduation. But now your parents have given you an opportunity to actually do something with your knowledge—and, incidentally, pay them back a little for the help they've given you with your education.

Your parents have decided to build a new house on the northeast side of the city. They've decided to make it not only attractive but also energy efficient, using the best insulation and design specifications available. They would also like an energy-efficient heating and cooling system, and that's where you come in. They've heard of solar power, of course, and a few other alternatives, but they don't know which type of system would be most economical and do the best job for their purposes. They ask you for suggestions.

As it happens, you have a suggestion for them—a heat pump system. You've just finished studying these designs for a class project and you think they might work well for your parents' purposes. After you describe the basic idea, your parents agree that it sounds promising. But they need more information about whether this system will actually work for them. You have to admit that you have little information about the specifics of the costs and payback time of the system. Your mother suggests that you do some research for them. The three of you agree on a plan: You'll research the system and write it up for them so that they can discuss installation with the contractor; they'll pay you for your time with a box of floppy disks (cheap—but these are your parents, after all!). As you get into your research you uncover these facts.

Notes

<u>Definition</u>: Pumps move heat from place to place. Examples: Refrigerators, air conditioners. Both pump heat outside. Heat pump can both heat and cool-- summer, heat goes out; winter, heat comes in.

<u>Ground-coupled heat pump</u>: Uses ground for heat extrac- tion and dissipation.

Efficiency measured by coefficient of performance (COP) ratios. Divide energy output by energy input. Heat pump COP: between 2 and 3; for every unit used to pump heat, 2 to 3 units of heat produced. Electric heat COP: 1; oil or gas heat COP: less than 1.

<u>Air-coupled heat pump</u>: Heat extracted from outside air and pumped into outside air. Problems: In winter heat has to be removed from cold air and moved in; in sum- mer, hot air has to be moved outside into hotter air. Bigger the difference in inside and outside air tem- perature, longer pump has to run to do job.

<u>Ground-coupled design</u>: Pump connected to plastic tub- ing buried underground. Tubing filled with antifreeze. Antifreeze circulated through tubing: transfer medium between pump and ground. Solar heat stored in ground

absorbed into antifreeze, transferred to pump. Reverse: Antifreeze carries heat from pump into ground.

Soil has less temperature fluctuation than air; warmer in winter, cooler in summer.

Full name: Closed-loop, ground-coupled heat pump system--CLGC heat pump.

CLGC components: Ground coil buried in ground; liquid-to-liquid evaporator/condenser; freon gas accumulator and compressor; liquid-to-air evaporator/condenser.

Accumulator/compressor similar to air conditioner--reversing valve added. Valve directs freon flow. Freon stored in accumulator, compressed to liquid in compressor.

Ground coil is buried closed loop of polyethylene tubing (11 1/2 in.). Length depends on system: usually 1,000-2,000 ft. Filled with antifreeze solution--same as car's antifreeze. Pump connected to coil, circulates solution between coil and liquid-to-liquid condenser.

Liquid-to-liquid evaporator/condenser--link between antifreeze and freon gas. Network of freon tubes in antifreeze allows heat transfer from freon to antifreeze. Freon lines connected to evaporator/condenser make for flow of freon to accumulator/compressor.

Liquid-to-air evaporator/condenser transfers heat between circulating freon and household forced air ductwork. Linked to accumulator/compressor by freon lines. Freon flows through network of coils--like radiator in

car. Fan blows house air through coils; heat transferred. Air distributed through house by conventional
forced air duct system.

<u>Heating</u>: Antifreeze in coil absorbs heat from soil,
then transfers heat to freon in liquid-to-liquid
evaporator/condenser (e/c). Freon boils into gas; gas
flows into accumulator and compressor--compressed back
to liquid (creates heat). Hot freon flows through
liquid-to-air e/c. Air blown by fan absorbs heat,
distributed through house.

<u>Cooling</u>: Freon flows through liquid-to-air e/c, picks
up heat from inside air, boils into gas. Gas flows
through accumulator/compressor, compressed into liquid.
Hot liquid flows through liquid-to-liquid e/c, transfers heat to antifreeze; antifreeze flows through coil,
transfers heat to soil.

Heating and cooling modes basically the same; change in
freon flow and thus flow of heat.

<u>Loop options</u>: Vertical or horizontal.

<u>Horizontal</u>: 60-ft trench dug around house; tubing
buried in trench. Good for large yard without obstructions (i.e., trees, walks, driveways, etc.). Requires
space for 1,000-2,000 ft of tubing.

<u>Vertical</u>: Hole drilled 200 ft to 400 ft deep (4 in.
diameter). Two lengths of tubing forced down to bottom
(tubing connected by U-shaped fitting). Good for small
yards or yard with lots of obstructions. Less excava-
tion needed.

Vertical system can sometimes reach below water table;
better heat transfer ability. Heat removed during
winter--ground gets colder around coil. Thus in summer, ground very cold, heat dissipation more efficient.

Soil will be dried out if too much heat is put into it
by horizontal system; heating load needs to be higher
than cooling load. Dry soil will mean gap between
tubing and soil; heat transfer efficiency cut. If soil
moist during summer (e.g., high humidity, good rain),
horizontal system will be OK. If heating/cooling loads
equal, ground around vertical coil will return to original temperature; if either load is heavier, there may
be problems.

Construction: PVC pipes and standard fittings unacceptable: leaks develop. System requires special
polyethylene pipe; uses butt-fusion device to eliminate
leaky fittings. 50-year warranty. Remainder of system
like regular heat pumps and air conditioners--should
have some reliability.

System can work in most climates, most areas. Main
requirement is contractor who is familiar with system
and uses proper system size, design, and installation.

Initial cost: Between $5,000 and $6,000, around $2,000
higher than other systems. Cost depends on site--
climate, soil type, heating/cooling loads. Biggest
cost is ground coil: $1.00 to $1.50 per foot, installed.

Operating costs: By my calculations, CLGC system would
cost around $250/year for heating--33% as much as an
electric system, 43% as much as an oil system, 47% as

much as a natural gas system. Air conditioning would cost around $200/year, approximately 35% as much as a conventional system.

Payback would take 1/10 the time of an oil and central air system, 2/10 the time of a gas or an electric and central air system.

Assignment 1. Find out whether any contractors in your area actually install ground-coupled heat pumps. Research local installers of heating and cooling systems; find out what the cost would be for installation. (Your parents have a quarter acre lot with no major obstructions.) Write a short informational report.

Assignment 2. Write the feasibility study for your parents. Your purpose is to explain what a ground-coupled heat pump is and whether it's a viable option for your parents' new home. Neither of your parents is particularly knowledgeable about energy systems.

Assignment 3. You decide to make use of the material you've collected for your class. Write another feasibility study on using the ground-coupled heat pump for your parents' new home; this time your audience is your professor.

INSTRUCTIONS AND MANUALS

Writing instructions or manuals requires the ability to anticipate the difficulties a beginner will face. As a manual writer, you also need the ability to visualize what might go wrong with a mechanism or a process. Finally, you need to be able to plan your instructions so that tips, warnings, and helpful hints are available where they are convenient. For these reasons, manual writing requires the same kind of mapping or flow charting skills demanded by computer programming or minimal path analysis. This introductory section first explains the preliminary stages of manual writing: classifying parts and describing parts and processes. After that, you'll learn about the instructional aids that help a manual to *teach,* not just describe or provide orientation.

AUDIENCE AND PURPOSE

The key to writing any manual is a clear sense of your audience (or audiences) and the purposes you intend to achieve. Do you need to provide manufacturing specifications? Assembly information for trained technicians? Troubleshooting or repair advice for general consumers? Maintenance data? As you learned in Chapter 1, your initial decisions about audience and purpose will dictate the scope of your manual and the language (technical or not), graphics, organization, and level of instruction. When you have a clear idea of audience and purpose, you are ready to approach the basic tasks of defining the subject of your manual for your audience and describing mechanisms and processes.

DESCRIBING MECHANISMS AND PROCESSES

Describing a mechanism or a process calls for a divide-and-conquer approach that includes four steps: defining the mechanism or process, naming the parts, classifying the parts, and writing the part-by-part description.

Defining

You can define a mechanism or process either formally or by comparison. Formal definitions follow a standard formula:

species = genus + differentia

or

thing to be defined = the class to which it belongs + a sufficient set of characteristics to distinguish it from other members of the class.

For example, a framing hammer is a hammer with a 16-ounce head and a claw for drawing nails. Less formally, one might define a framing hammer by calling it "the hammer commonly used in carpentry." A definition by analogy might be "a Kaypro 2 computer is like an Osborne 1, except that a Kaypro has a full width 9-inch screen and two disk drives." You will need a definition if your audience is not familiar with the process or product you are describing.

Naming Parts

Next, name the parts. This step is important. One approach is simply to number all the parts. This approach works well if the parts are difficult to name or have names that are long and confusing. Numbers, however, are not memorable. If you choose to name parts, select the names carefully so that they can be used without confusion in your description. One student described a stapler by assigning the names shown in Figure 5-1 to the principal parts. As you can see, using the words "top" and "bottom" repeatedly leads to total confusion in a paragraph describing the operation of the stapler.

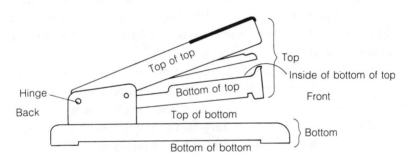

Figure 5-1. The wrong way to name parts

Classifying Parts

Third, classify the parts into groups. Manual writers and others who write descriptions often make the error of numbering parts clockwise around a mechanism, forgetting that parts are grouped naturally into subassemblies. When you have developed the best way to divide your mechanism or process for easy understanding, be sure that you have included those small parts that are easy to overlook. For example, model kits for children often forget to mention the type of glue that is needed.

Writing the Description

When you have taken inventory of the entire process and clearly named all the steps, parts, and equipment, then you can write the part-by-part or step-by-step description. Start by providing an overview of all the main parts or steps and how they fit together in sequence. If you provide a figure or illustration, it is common to describe parts or steps from the left to the right of the illustration. Next, take the first part or step and provide an overview of the subassemblies; then describe each subassembly in turn, discussing sizes, shapes, materials, finishes, colors, and function. At the end of each subsection of your description, describe a cycle of operation. At the end of the entire description, recap with a cycle of operation of the complete mechanism or process. As the outline in Figure 5-2 illustrates, the description consists of envelopes within envelopes of description, each unit "glued" to the next by a cycle of operation.

Throughout this brief introduction, we have used spatial order in our examples. For processes, temporal order is often more appropriate, and introductory sections of manuals should generally be organized according to the importance of the information. This subject is covered in the next sections.

ORGANIZING YOUR MANUAL

Description of mechanisms and processes are usually only part of some larger document, such as a manual. A description by itself is not an instructional, installation, or maintenance manual because a manual needs to teach, not merely list. If you plan to develop a manual from a description of a mechanism or process, at a minimum you need to include (1) a list of tools and materials that may be needed, (2) performance instructions, (3) performance criteria that will tell a user when a step has been completed

```
1.  Outline of a description of a mechanism or
    process
         a.  Overview of parts or steps
         b.  First part or step
              (1)  First subpart or substep
                     (a)  shape
                     (b)  location
                     (c)  material
                     (d)  etc.
              (2)  Second subpart or substep
                            .
                            .
                            .
              (3)  Cycle of operation

2.  Second part...
       .
       .
       .

3.  Complete cycle of operation of process or
    mechanism.
```

Figure 5-2. Sample outline

properly, (4) troubleshooting assistance with common problems, (5) cautions where danger to the equipment is possible, and (6) warnings when personnel may be endangered by a malfunction. All of these user aids should be tailored to the needs of the audience. Moreover, you should flow chart your entire manual before you begin writing to assure that troubleshooting sections, alternative paths, and related materials are clearly connected by a network of signposts that any reader can follow. The following outline suggests a useful order for the main sections and lists the teaching aids that should appear wherever necessary throughout the manual:

1. Overview, purpose, and scope
2. Principles of operation (hows and whys that are not familiar to your audience)
3. Tools and materials
4. Descriptions of mechanisms or processes
5. Teaching aids
   a. Warnings of danger to personnel
   b. Cautions of danger to equipment
   c. Notes referring to contingencies, including common problems of operation or maintenance
   d. Performance criteria (What should happen if you are doing it right?)
   e. Troubleshooting help or a troubleshooting appendix
   f. Visual aids
   g. Teaching sections that provide useful tips from the perspective of the absolute novice

WRITING TO INSTRUCT

Above all else, manuals need to teach *how* to do tasks. The following features are hallmarks of manuals that teach effectively.

1. Use imperative verbs. "Insert the handle into the left side, twisting gently . . . ," *not* "The handle is inserted into. . . ."
2. Write steps that are gramatically parallel. *"Open* the hood, *remove* the wingnut from the air filter, *pull off* the air filter cover, *inspect* the carburator for dark deposits."
3. Arrange steps in columns so that they are easy to follow, and use numbers and bullets to distinguish one step from the next.
4. Group steps that are part of one stage of an operation. For example, group disassembly steps in one area (visually and by means of organization), and then establish separate groups for repair procedures and reassembly procedures.
5. Put no more than seven steps in any one group.
6. Place graphics and warnings near the place where they are first mentioned, and use graphics and warning statements to separate major groups of steps.

CASES

THE PUBLISHING PROGRAM

Your employer, Apollo Software, has just developed what it considers to be a revolutionary program for electronic publishing. The program, Page Performance, will create document layouts, including setting up columns, placing graphics, inserting headlines and subheads of differing sizes, and even adding design elements such as borders and screens. Using this program a small publisher can design a newsletter, brochure, or even a book entirely on a computer console. It is obvious that these capabilities will provide a major advantage for those who use the program, but it is equally obvious that the program itself is extremely complex. That's where you come in.

As a member of the technical writing staff at Apollo, your job is to create a user's manual that will help those who purchase

Page Performance to make their way through the maze of commands and options. The head of the writing staff has decided that one way to help the customers become comfortable with the program is to begin the manual with a practice session—nothing fancy, just a quick run through some of the standard commands and operations the program provides. You and your coworkers have been trying out the program for the past two weeks, keeping track of the procedures you used with the commands and the problems you encountered as you tried to use the program. You've been concentrating on writing the operations necessary to create the column layout and place text and graphics in the columns. Another writer, Jeanne Cooper, has been going through the program to see how it works. She has given you a rough description of her experiences.

```
TO:  Cole Younger
FROM:  Jeanne Cooper
SUBJECT:  Creating Columns and Placing Text/Graphics
with Page Performance
```

Frankly this thing is so full of bugs they should use insecticide on it, but the following is roughly what should happen if everything is working well.

When you get into the program (assuming they get the bugs out and the thing boots up the way it's supposed to) you see the page on the screen. That's the reduced size view because it shows you the whole page at once. If you want to see the thing in real size (i.e., section of page you want to work on), you have to simultaneously hit COMMAND and P. That highlights the Page menu; then you have to use the cursor keys to move the cursor down through the selections on the menu until you highlight "Actual." Then, press the ENTER key.

That makes the page the actual size on the computer so you can see the words you're putting in.

Before you do that, though, you should set up the basic layout of the page you want to use. While you're on the reduced size view, there's a series of three boxes in the lower left-hand corner of the page. One says M, one says L, one says R. L and R are the left and right pages of a doubled-sided publication. If it's single sided you just see R. (You choose single-sided or double-sided back with the initial commands; that's the section Maria's working on so you'll need to check her memo.) M stands for master page. If you use that you can set up a layout that will apply to every page of the document. To make a master page you press COMMAND and P. That puts you into the Page menu again and you highlight "Master" with the cursor and press ENTER. That highlights that M box on the screen, too. After that anything you put on the page in the screen will be placed on the master page. Whatever box is highlighted on the screen, that's the page you see.

If you need a ruler for setting things up you press COMMAND and T. That opens the Tools menu. Then you move the cursor to highlight "Ruler" and press ENTER and that puts a ruler on the page.

To set up the columns on the master page or the other pages you hit COMMAND and T. That's the Tools menu again. Move the cursor to "Columns" and press ENTER. That opens up the Columns menu; you can enter the number of columns and the width of the space between the columns (the "gutter"). N.B. That sets up columns on whatever page is highlighted. After you enter the

numbers you move the cursor to OK on the Columns menu
and that closes the menu box.

To place text on the page: First of all, be _sure_ that
you're off the master page and on page 1. Hit COMMAND
and P (Page menu), highlight "First Page," and press
ENTER. Check to make sure that the L box is highlighted down in the lower left corner. (If it isn't
you're not on page 1 yet and you have to do the whole
thing again.) It's not a good idea to actually type
the text onto the page. The word processing commands
are _very_ limited and it'll take you twice as long. You
use the word processing strictly to correct errors.
Type your text using another word processing program.
(Maria's got the list of the programs that are com-
patible--most of them are.) You put the disk with the
text on it into drive #2 (Kevin's covering all of
this). Next you pull down the Text menu (COMMAND and
L) and highlight "Place" with the cursor and press
ENTER. That opens another menu (the Place menu) that
lists all of the text files on the word processing disk
in drive #2. You highlight the title of the text file
you want to place and press ENTER. This loads the file
into the memory. The menu should disappear. Now this
gets tricky. You have to place the cursor exactly
where you want the text to start on the page. I think
it's easier to do this if the screen shows the page on
actual size, but you can do it either on actual or on
reduced size. When you've got the cursor where you
want it you hit COMMAND and Q. The text _should_ be
placed on the page then. It will flow into whatever
column spaces you've already set up.

To place graphics: You can place graphics either before or after you place text; it'll work either way.
Basically you've got to have the graphics on a disk

that you put in drive 2 (just like the text). You
highlight the Graphics menu with COMMAND and G. Then
you highlight the title of the graphic file you want to
place and press ENTER. Then basically you place it the
same way you place text--put the cursor where you want
the graphic to be and hit COMMAND and Q.

If you've already got text on the page when you place
the graphic, you'll want to flow the text around the
graphic. If you put a graphic onto a page with text,
the text will flow <u>over</u> the graphic, but you can see
the graphic beneath the text so you'll know where to
break the text line. Move the cursor to the point on
the text line where you want to break the line (usually
right next to the outer edge of the graphic), then hit
RETURN. Do the same thing for each line until the
graphic is clear of the text. The text <u>should</u> realign
and flow between the column guides on the page. Incidentally you can place graphics outside the columns
you've already set up; that is, you can have a graphic
that spreads across two columns or one and a half
columns and so on. However, you can't place text
outside the column guides without going through another
series of commands which I think are too complicated to
put into the practice session. (It's called the Column
Override Command Sequence if you want to put in a
reference to it.)

Assignment 1. Write out the material Jeanne has given you in the
form of a series of instructions. These instructions are largely for
your own benefit and for the benefit of the others in the writing
group. You need to have a set of instructions to follow when
you're working with the program before you write the actual
manual. Write the instructions for the group. All of you in the
group are experienced word processor users.

Assignment 2. Write the practice session for the manual. The practice session you are writing will be placed first in the manual before the actual instructions for using the program. The program will come with a practice disk complete with a text file (titled Practice Text; the text from the file will fit perfectly into a 3-column format with a 0.25-inch gutter) and a graphic file (titled Practice Graph; the graphic from the file measures 1½ columns). You are to write the part of the practice session that takes the user through the commands for setting up columns and placing the practice text and graphic. The practice session should consist of instructions and descriptions of what will happen on the screen when the instructions are followed correctly. Also indicate what pictures should be included in the practice session and where they should be placed. Because this is a sophisticated program, your users will probably have some experience with computers and word processing.

   The following sample of the opening part of the practice session was written by another member of the group. Use this format if you wish; however, since the project is still in the formative stages, your group leader is allowing each writer to experiment; you may develop your own format if you prefer.

```
Your first step in using Page Performance will be to
load the program disk and the special practice disk
provided with your copy.
•Put the program disk into drive A of your system (the
drive on the left) and close the door of the drive.
•Put the practice disk into drive B of your system (the
drive on the right) and close the door of the drive.
Your screen should display the words "Page Performance,
Apollo Software, Copyright 1988" in the upper right
corner.
              [Insert drawing of screen
                  showing message]
Your next step in setting up a new document is to
```

```
create a master page, a page that will contain any
layout features you intend to use throughout the docu-
ment, such as columns, borders, page numbers, or
titles.  To set up your master page
•Press the COMMAND key.

Your screen should now show a display of five words
across the top:  Page, Tools, Text, Graphic, and Edit.
                [Insert drawing of screen
                   showing menu titles]
•Press the cursor key until "Page" is highlighted.

Your screen will now show the Page menu.
                [Insert drawing of screen with
                      Page menu]
•Press the cursor key until you highlight "Master" and
press ENTER.

Your screen should now show a display of the master
page.
                [Insert drawing of screen
                   showing master page]
```

THE MULTI-CULTURAL TRAINING MANUAL

Now that Hadley Engineering is winning contracts outside the United States, the company is experiencing first hand the problems of hosting foreign guests and preparing staff engineers and their families to live abroad. As training manager, you know that many large corporations like Boeing Aircraft maintain divisions that train and house foreign pilots, teach languages, and help sales staff and engineers to understand the social conventions and business practices of other cultures.

Hadley cannot invest in a major corporate educational center right now, and with engineers and their families leaving for foreign cities on an unpredictable schedule, hiring an expensive outside training firm to offer seminars is not practical.

You have been asked, therefore, to write several short manuals that will be given to employees who are going (with or without family) to *specific* countries. The manuals must be short—fewer than six pages of copy—and carefully organized to emphasize the main points. Hadley has tried giving away histories of nations and semischolarly studies of other cultures and other religious traditions, but people who are moving 10,000 miles don't generally make time to read long books.

Your boss wants you to start by producing *one* manual, choosing any one of the following nations:

Kuwait

Saudi Arabia

Jordan

Taiwan

Egypt

Thailand

Indonesia

Peru

Singapore

Angola

Romania

Poland

France

Korea

People's Republic of China

Venezuela

Nigeria

The Philippines

Tanzania

Your boss reminds you that what holds true for one Arab nation (for example) is not necessarily true for others, just as there are differences between the United States and Britain or any other nations that share the same general cultural or historical traditions.

As a beginning, your boss offers you this list of areas to consider:

legal status of foreign nationals

political system

absolute taboos

talking to strangers

languages

living conditions: water, food, marketing, health conditions

eating meals: women, children, styles

customs relating to handshakes, visibility of feet

religion, attitudes toward other faiths

women's status and women's dress

expectations for dress of foreign women

men's dress and expectations for foreigners

exchanges of gifts, expectations of gifts or "bribes"

tipping

patterns of greeting

negotiating styles

attitudes toward time and pressure in negotiations

Your boss then explains that he realizes that these issues will require different emphasis in any particular manual. Moreover, in some cases, some of these areas will not need to be addressed while others will be very important.

Assignment 1. Prepare *one* manual, using a combination of library research and personal interviews to gather data. Provide your instructor with the names of people you interview in your research. Keep in mind that most foreign nationals living in the United States will appreciate an opportunity to explain their positions and cultures as long as the American interviewer is sensitive, willing to listen, and nonjudgmental.

Assignment 2. Sixteen executives are being sent to the country you chose to research. Your boss wants you to give a twenty-minute oral presentation at their last sales briefing. Your manual will be passed out after the presentation. In your talk, therefore, you want to focus on (1) the most important lessons and (2) the information the executives will need as soon as they get off the plane. Design visuals to support your oral presentation.

THE LASER PRINTER

As a technical writer for Elgin Electronics, you have worked as an assistant on several projects. Now the time has come for you to write something on your own. Elgin's newest product is a laser printer for office use; your job is to write the maintenance section of the user's guide to accompany the printer. (Other writers in your group are writing other sections.) For the past three weeks you've been assembling materials and now you need to review your notes.

First there is the basic information you got about the printer from interviewing people in the engineering division.

Notes
    Elgin 734 Laser printer
For office applications: High-speed, high-quality
laser-scanned electrophotographic printer
Controls: <u>Pause</u>, <u>Run</u>, <u>Power</u>. Also a printer lock on
the front panel. Pause and Power on upper left, lock
lower center. Other controls available if they use
computer network options--User's Guide will give details.
Lock: Key operated. Controls front panel. Front
panel must be opened to get to paper feed and engine.
Pause button must be pressed before panel unlocked.
Power: Down for off, up for on. Internal circuit
breaker; in case of power loss, push switch off and on
again to reset.
Run: Starts printing, locks all access doors.
Pause: If you want to stop printing, press pause. 734
keeps going until pages being done are complete. Pause
flashes after button pressed. Doors will release;
pause will light up (not flashing). You have to press
Pause to add paper or toner--any time doors or drawers
opened. Press Run to get things started again. Pause
and Run both flash for paper jam--continues until jam
cleared and doors closed.

Next, you need to know about adding paper and toner. Carol
Sutorius, one of the other writers in your division, was originally
working on the maintenance manual but was reassigned to another
job. You speak to her about these two routines.

<u>Interviewer</u>: Carol, Sean in engineering said the 734
    uses paper drawers; how does that work?
<u>Carol</u>: Basically you've got three different input
    paper drawers or bins that contain movable trays
    that you can load. Each holds up to 500 sheets.
<u>Int</u>: Any particular paper weights or sizes?

Ca: Not really. They'll handle just about any weight or size, including transparencies. Tell them to check the User's Guide if they have any questions.

Int: So how do they add the paper?

Ca: OK. First you have to press Pause. You always press Pause, no matter what you're going to do to the printer; that's always the first step. Also they have to wait until the light glows.

Int: What, you mean the Pause light?

Ca: Yeah. Wait until it stops flashing. Then they take out the paper drawer that has the coded tray for the type of paper they want to use.

Int: Wait a minute, you've lost me! Coded trays? What do you mean?

Ca: OK. Each of those trays has a coded sensor that goes with a particular type of paper so the printer can take the right paper from the right tray. You need to get the right paper in the right tray. They get eight trays with the printer, each coded for a different type of paper; they store the five that aren't being used.

Int: So how do they find the right tray for the paper they want to use?

Ca: Check the User's Guide. It has a table of all the various paper types and sizes.

Int: All right, go on.

Ca: Then you put the paper in the right tray. The trays have lines inside them with size indications on them. They just line up the size paper they have with the right mark. Then they put the tray back into the input drawer and put the drawer back. Those drawers are inside the right end of the printer.

Int: So they have to unlock the door to get at them.

Ca: Right. Always. Then press Run to get things
going again.

Int: Will the printer run with the door open?

Ca: No. The door has to be closed so the door locks
can reset when they push Run.

Int: How about letterheads, paper with stuff printed
on it?

Ca: Well, the main thing is they have to put the side
they want to print on--the side with the letterhead
or whatever--face down in the tray. The top goes
toward the back of the tray. And, of course, they
have to have the right type of tray for that type of
paper.

Int: How about double-sided printing?

Ca: Just take the paper out and turn it over--printed
side up, top pointing toward the front of the tray.

Int: What about adding toner?

Ca: They've got to use our toner; the number is 1830.
If they use anything else they can mess up the
printer. The console screen tells them when the
toner is low.

Int: How?

Ca: It flashes Add Toner

Int: Catchy. Then what.

Ca: OK. Press Pause, right? Then open the front
panel. There's a lever right in front that's
labelled Toner Release; they push that to the right.
That releases the toner drum; you can just pull it
out. It's right behind the release lever. They
take off the top and pour in the toner; then they
put the top back on and put the drum back behind the
lever and push the lever back the way it was before.
Sometimes, if they haven't pushed the drum all the
way in, you can't get the lever to slide back. So

if they're having trouble getting the lever back in
place, that's probably it. Incidentally that toner
is really messy stuff. They should put that drum on
something that can be cleaned up easily when they're
through filling it--not a carpet or somebody's lap!
After they've got the drum back in, they push the
door closed and push Run and that's it.

Next you need to know about clearing paper jams. The engineering department has already written a preliminary version of
these instructions and given you a copy.

Paper Jam Procedure
    The 734 is a high-speed, multiapplication precision
printer offering multi-input drawer configuration and
high variability in paper option capabilities. However, high speed means paper can become lodged in
copious locations within the printer configuration.
Jams can happen in the infeed path, the outfeed path
and the electrophotographic engine region. Printer
capabilities mean that console locator guide will flash
PAPER JAM AT. . . and give the location.
    The general clearing procedure is as follows.
Pause button should be pressed to stop any printing
action. Front panel unlock should be effected and
opening should follow. Having been given the location
on console locator guide, location should be checked
and paper cleared from area. After clearing, trays and
latches should be returned to original positions.
Close and press Run. If jam is at the printing drum,
toner release lever should be pressed. Then remove
toner drum--be careful because it can leak toner and
soil clothes. Pull out printing drum release lever and
printing drum will be accessed. Drum should be slid
out <u>very carefully</u> because it can be scratched, dented,
or soiled easily and these drums are very expensive to

replace. No one should touch the surface of the drum
ever. Paper jam can now be cleared. Reverse process
to get drum back in place.

So far so good, but as you go over your notes you realize
there's one area you haven't covered: removing copies from the
printer. You finally run down Sean Copeland, your contact in the
engineering department, to get the information. Sean is in the
middle of another job, but he agrees to answer a few questions.

Interviewer: Sean, what happens to the copies when
    they're printed?
Sean: Up the outfeed path and into the bins--6 bins,
    that's the output distributor. The user can desig-
    nate the bins for each paper type.
Int: Does the User's Guide cover designating the bins?
Sean: Yeah, I guess so.
Int: So how do they get the copies?
Sean: Open the door and take the stuff out of the
    bins. They're on the left side.
Int: Do they have to press Pause first?
Sean: Yeah.
Int: And Run to start again?
Sean: Yeah, yeah.
Int: What about this lockbox thing?
Sean: OK. That's for confidential stuff, things they
    want to print that they don't want anybody to see.
    If they use one, they need two keys. One gets the
    lockbox out of the output tray and the other one
    unlocks the box.
Int: So it's press Pause...
Sean: Press Pause, unlock lockbox from output tray,
    pull out lockbox, unlock it, take the stuff out.
    They just reverse it to put the thing back in.
Int: Wait a minute. You mean they put the tray into
    the bin and then lock it in with a key?

Sean: No, no. Just slide the box into the bin until
they hear it click in place, then it´s in.
Int: Sounds simple enough.
Sean: Yeah, hey, wait though. There are some things
they shouldn´t do with those boxes. They shouldn´t
leave them in the bins unless they´re locked in and
they can´t put them back if they´ve removed them
unless they´ve been unlocked with the second key.
Int: You mean the box has to be unlocked and emptied
before they can put it back?
Sean: Yeah. And they´ve always got to hit Pause
before they put the lockbox in the bin.
Int: Thanks.
Sean: Any time.

Assignment 1. Now that you've got your materials assembled, it's time to set up the graphics. Write a memo to the graphics department describing the graphics you want them to design for this manual.

Assignment 2. Write the maintenance section of the User's Guide for the Elgin 734 Laser Printer. Indicate where you would insert graphics and what those graphics would consist of.

THE EDITING JOB

Few writing tasks are more challenging than writing descriptions of complex physical movements in three-dimensional space. To simplify such descriptions, manual writers learn to group related steps, to insert frequent subtitles, and to obey the "rule of seven." This rule states that no more than seven steps should appear in a single group. Beyond that number it becomes difficult for readers to comprehend a group of steps as a single unit.

It is possible that some physical movements and procedures can only be learned by watching someone do them or by trial and error. Nevertheless, writers must often do their best to instruct audiences in complex maneuvers.

For this case, you must first choose a maneuver to study and then find at least two written (preferably published) descriptions of that maneuver. You may wish to consult with your instructor or other students as you make your choice. Some possibilities include:

1. Paddle strokes in canoeing or kayaking
2. The Pawlata roll or Eskimo roll in kayaking (or some other rescue procedure in boating)
3. A skiing technique
4. A description of a movement in gymnastics
5. A movement studied by kinesiologists
6. A surgical procedure

Although it might be to your advantage if you know how to perform the movement or procedure yourself, this is not necessary. Technical editors, after all, must often edit and improve texts that describe actions and information they do not fully understand.

Assignment 1. In a short report to your instructor, compare the two descriptions of the maneuver you have chosen to study. Which author does a better job of describing and teaching, a better job of manual writing? Compare not only the words but the layout, the use of subtitles, and the graphic support.

Assignment 2. Edit one of the descriptions you have found. Consider audience and purpose. Has the original author written to experts rather than to novices? Can technical language be cut or simplified? Can steps be grouped or set apart on the page so that the total process is easier to comprehend and perform? Provide your instructor with a xerox copy of the original and your own edited version. Because, for the purposes of this assignment, you are imitating the work of a professional editor, you are allowed to quote substantial sections of the original and use graphics from your source. However, you should be careful to avoid using materials that should be rewritten as part of your editing job.

Chapter **6**

PROPOSALS

Proposals are sales documents. They are intended to persuade someone to issue a contract. But proposal writing is also a political process in which the written proposal often serves as a record of terms that have already been agreed to in conversation. Proposals normally address four questions: what will be done, how much will it cost, who will perform and supervise the work, and how will the results be evaluated. These questions are often answered in four separate sections called the *technical, cost, management,* and *evaluation* proposals.

THE PROPOSAL PROCESS

The proposal process starts either with an idea or with an opportunity. If you receive a request for a proposal from a corporation or government agency, you have an opportunity to respond with what is called a *solicited proposal.* If you decide to send a proposal on your own initiative, it is called an *unsolicited proposal.* Proposals often start with an *abstract,* a two-page discussion of the general approach to a problem. Proposal abstracts are distributed within a company so that interested persons can comment, join the project, or offer support. After the abstract has been reviewed, in many agencies a proposal project is taken over by a central proposal office. The function of a proposal office is to select the best target for a proposal and to initiate communication with the target agency. Proposal negotiations often include phone calls, visits, bidder's conferences (where all bidders on a project are invited to come and ask questions), and the exchange of abstracts and prospectuses. A *proposal prospectus* is a five- or ten-page executive summary of a full proposal. While abstracts are often limited to a discussion of technical and management questions with less emphasis on costs and evaluation, a prospectus tries to address all four proposal questions. Small projects may be funded on the basis of the prospectus, which can also be called a *short proposal.* Short proposals can be in letter form. If a longer proposal is called for, it normally takes one of two forms: the multipart proposal or the narrative proposal.

MULTIPART PROPOSALS

Multipart proposals are used in highly technical fields. Figure 6-1 shows an outline of a multipart proposal.

Some outlines of multipart proposals contain many more sections than the sample in Figure 6-1. Typically, full-scale proposals include management, cost, and evaluation sections. The management proposal persuades the buyer that your company is qualified to do the work because it is well managed, staffed by qualified personnel, and has a history of doing similar work. Cost proposals and budgets include cost estimates and a summary of

1. Front matter (cover, letter of transmittal, overview or executive summary of
   entire proposal)

2. Technical proposal
   a. Statement of problem
   b. Proposed solution
   c. Feasibility of proposed solution
   d. Conclusion
   e. Appendices with supporting data

3. Management proposal
   a. Project management
   b. Company history

4. Cost proposal

5. Evaluation proposal

Figure 6-1. Sample outline of a multipart proposal

the methods used to reach them, particularly for estimates that might be controversial. Evaluation proposals are most common in social service and education projects, where the product is difficult to measure with precision. How does one measure, for example, the effectiveness of a community education program against child abuse? In such a case, proposal writers try to develop an evaluation plan that combines internal evaluation of a program by the staff with evaluation by external (or visiting) experts. In addition, there is usually an attempt to combine objective criteria, such as the number of law cases, with subjective criteria, such as the opinions of qualified experts in the field.

NARRATIVE PROPOSALS

While a proposal to build an aircraft demands a multipart proposal, a proposal to reduce child abuse through community education calls for the narrative format. In the narrative proposal format, technical, management, and evaluation material are combined into a running narrative. Often the narrative provides historical context for the current proposal. For example, if you wanted to establish a program of television advertising, clinics, and workshops to deal with child abuse, you might discuss public interest in the issue, media coverage, methodologies and approaches recommended in the professional literature, evidence of the effectiveness of clinics and media campaigns in other cities or countries, and specific objectives of your proposed program. As you can see, that list of issues broadly combines technical, management, and evaluation material. The danger implicit in the narrative format is that the individual management and evaluation material will not be isolated for careful discussion and presentation. One solution to this problem is to follow the historical narrative with short subsections on each of the four main proposal sections.

As you respond to the case assignments in this chapter, use the multipart proposal outline to guide your thinking about audience, purpose, data gathering, and problem solving. You will also find it helpful to reread the sections on recommendation reports and feasibility studies in Chapter 4. The proposal is closely related to these reports; the recommendation report work sheet and the information about formulating and using criteria are useful in writing proposals as well as recommendations.

CASES

THE NATIONAL INSTITUTE FOR RESEARCH

Many research grants are available to you as an undergraduate. This assignment asks you to prepare a research prospectus according to guidelines similar to those published by many foundations, universities, and government research programs. This assignment offers you a chance to develop skills you may wish to use in applying for actual grants in the future. By completing this assignment, you will learn more about your own field of study.

<u>National Institute for Research</u>
Junior Researcher Program

The National Institute for Research offers summer research grants in the amount of $2,500 to qualified undergraduates whose proposals meet the requirements that follow. Up to 50 grants are offered each year on a competitive basis. Grants require two months of full time work on a carefully defined project under the direction of a supervisor who has the qualifications to advise and evaluate the project. Examples of projects include research on a plant or animal species (such as

a sea grass, a protozoan, a marine parasite), development of an educational program, research on an issue in computer hardware or software, historical research in regional history or literature, or an engineering design project. Any project that is judged appropriate for an honors thesis at the bachelor's level or for a two-term independent study project might qualify.
Every student participant must prepare a research monograph describing the work in detail, the results of the work, and the sources (bibliography, research contacts, facilities) used in the course of the project.
Eligibility: Applicants must be 23 years old or younger. Students over 23 may apply if they are enrolled full time in an undergraduate program and completing their first baccalaureate degree.

Proposals must call for work by a single student researcher. No joint or group projects will be funded under this program.
Project Categories: Proposal projects must fall into one of the following three categories.
1. Broad study of current work in a subspecialty within the student's major field. Applicants may study major theories, current research foci, and/or the work being done at a major research center.
2. Advanced study of a specific research area in the student's field, with a specific and narrowly defined product, goal, or hypothesis to be tested.
3. Interdisciplinary study focusing on technology or theory transfer from one discipline to another (such as applications of laser technology to the study of molecular structure, computer analysis of organizational or management structures, applications of miniaturization and the manufacture of miniature devices to fields other than computer chip technology, the transfer of developmental psychology to the study of writing, or the application of neuroscience to linguistics).

<u>Evaluation</u> <u>Criteria</u>: Projects will be evaluated according to the following criteria.

1. Does the project fit into one of the three categories listed above?

2. Is the scope of the project well defined?

3. Can the goal be reached within the time limit (two months)?

4. Is the project well defined? Has the applicant prepared a step-by-step analysis? Is the project sufficiently rigorous and demanding?

5. Does the proposed advisor have the necessary qualifications to supervise and evaluate the research?

6. Does the project address an issue that is important in the discipline?

7. Is the applicant qualified (in terms of academic performance and previous coursework) to carry out this project?

<u>Application</u> <u>Process</u>: The application must contain the following sections.

1. Project definition. Describe the problem or issue or hypothesis to be examined during the period of the grant. Explain the questions that you will explore and explain your approach or methodology. Explain how the project fits into one of the three project categories. Explain how the project idea was developed--what its relation is to previous coursework or experiences. Explain how the project will contribute to your further education or professional objectives. <u>This work must be independent of any regular coursework or degree requirements</u>.

2. Describe in detail the research plan, including a proposed schedule for all work to be completed. Special attention should be given to the following areas: preliminary reading or study, meetings with the advisor, methods to be employed, necessary travel, schedule for testing any hypotheses, schedule for a written

report. If possible, this information should be integrated into a single, comprehensive, week-by-week
schedule.

3. Attach a list of references, collections, or laboratory facilities to be used, sites to be visited, and
probable contacts or interviews.

4. Describe your educational background, relevant
courses and independent study projects, career and
intellectual interests, project-related work experience, current academic status (year in college), and
year of anticipated graduation.

Length of Application: Sections 1 and 2 of the application (Project Description and Methodology) combined
should be no longer than six double-spaced typewritten
pages. The bibliography and list of other resources
should be a single page and should focus on only key
publications and resources.

Decisions of the judges are final. Keep copies of
all materials. No application materials will be returned.

Assignment 1. Research and write a project proposal. You are not
required to secure the assistance of a project advisor, although you
may wish to discuss a project opportunity with a faculty member from your own discipline.

Assignment 2. The National Institute for Research is willing to
consider applications submitted by a small group of students.
Write a proposal with a group of three to five students in your
own (or related) fields.

THE COMPUTER ROOM

Two years ago you accepted a position as head of data processing at the Coleman Research Institute, a private research foundation specializing in product testing. Most of your work has been routine, but you've been increasingly frustrated by one major problem: the institute's computer facilities. The rooms assigned for the computers are inadequate in a number of areas.

First, there's a problem with the air conditioning system. Since the Institute is located in a warm climate (average daily temperature of 70°F), year-round air conditioning is necessary. The entire Institute has a central air conditioning system, but on the hottest summer days this system has been ineffective in cooling the computer rooms. As a result, window air conditioners have been installed on four walls, but heat continues to be a problem.

The second problem is the floor. All of the computer equipment is interconnected by a series of cables. In other computer rooms you've been associated with, these cables are laid beneath a raised floor both to eliminate hazards and to make the room easier to move around in. The Institute's computer rooms, however, have no raised floor and the cables are a major annoyance.

Third, the only protection the rooms have against fire is hand-held fire extinguishers placed at intervals around the walls. These are adequate only if a fire occurs when the lab is occupied. Ideally the rooms should have an automatic system.

Fourth, the room should have a motor generator as an alternate power source during power fluctuations. As it is the system crashes periodically; data are lost and the users complain bitterly.

Finally, the room is simply too small and there is no room for expansion as the system grows. Although the Institute has no immediate plans to expand its computer facilities, you know from experience that such expansion is inevitable.

Taking all these things into account, you decide to investigate the possibility of a new computer room. Your plan is to assemble as much information as possible and then to write a proposal to the Institute's Board of Directors.

At this point you decide to do an equipment inventory so you can show exactly how much space is needed. The computer facility has two mainframes: a Digicom 500 and an RDS Excell. For the Digicom system you have a central processing unit (CPU), eight disk drives and controller, two tape drives and controller, a teletype, a card reader, a printer, and three communications devices. The RDS is currently housed in an adjoining room. It has a CPU, four disk drives and controller, two tape drives and controller, printer, microprocessor, teletype, and plotter. There is also a tape library split between the two rooms.

For the sake of convenience you would like to see both mainframes in the same room, which would be designed for the latest equipment.

Your first step is to interview the Institute's Director of Facilities, Joe Hasslocher. Joe has been at the Institute for more than twenty years and has an intimate knowledge of the background for each of the facilities. You discover that he's also very much in favor of a new computer facility.

> Interviewer: Joe, when did the Institute first set up those rooms?
>
> Joe: Oh gee, it's been over 25 years now, back in '60 or '61. We started out with a remote batch system, you know. Had a printer and a card reader and a teletype. Then we patched in to the main computer in California. We added a microprocessor about ten years ago. Then we got the Digicom in 1980. Of course, now the Digicom is linked into the remote system too so we've got both in that room.

Int: When did you get the RDS?

Joe: In 1982. The idea was for the Digicom to handle all the business administration work and some project work and for the RDS to handle just project work. We felt like we needed two mainframe processors; one just wasn't enough for what we were doing. That's when we added that adjoining room.

Int: So the Digicom room was originally set up just for the remote batch?

Joe: Right. The air conditioning was OK for that system. Then when we added the Digicom the room got too hot in the summer so we added the single units.

Int: But we still have some down time because of the heat in both rooms.

Joe: Right, I know. The RDS caused some more problems, and then the window units were never really the best solution, just sort of a stopgap kind of thing.

Int: What kind of service contract do we have with Digicom and RDS? Would they help us move the equipment and reinstall it if we had a new computer room built?

Joe: Our maintenance agreement covers that. Anything from 8 to 5, Monday through Saturday, the Digicom people will do. RDS has fewer hours on Saturday, 8 to 12, I think, but the agreement is roughly the same. In fact, that would probably be the best time to do it--on a Saturday when you've got minimal demands on the system.

Your next interview is with Luis Mendez, the Institute's architect. At this point you want to consider some alternatives for the new system.

Interviewer: Luis, what are the possibilities for expanding the data processing center? The place needs a complete renovation: new floor, better fire extinguishing system, new air conditioning, more room to expand. You know how bad it is now.

Luis: Yeah, I've heard the complaints. Well, as I see it we've got a couple of alternatives. We could renovate the room--add a new air conditioning system, raise the floor, put in a fire extinguishing system, all the rest of it. Or we could build an addition for the computer room with all of that stuff built in.

Int: Which do you think would be better?

Luis: They'd both work, but you've got to consider some other factors. If you look at just construction costs, the renovation would probably be cheaper, but consider this. If you want to raise the floor in the computer room, you have to move all the equipment out to do it; that means the computer would be down for several days and maybe longer. Same thing with the air conditioning; you couldn't leave the computer on while you were putting in a new system. There'd be too many problems with heat and humidity.

Int: And that room is too small as it is. Could the renovation enlarge it?

Luis: It could, but it would be expensive. We'd have to take out some walls. Plus that assumes that we've got the extra space, that we can relocate some other personnel so that you could add their space to the computer room. Believe me, that's not the case.

Int: So could building a new computer room help out some of the other departments? I mean, could they use the space the old computer room occupied?

Luis: Absolutely. For instance, you could move accounting into the old computer room with minimal renovation. They already have some space on that

floor; you could move the whole department there and
it'd be a lot more efficient.

Int: How would we go about building a new addition?
Would we have to hire a contractor?

Luis: For the actual building, yes. But our people
have the expertise to do the wiring and interior
carpentry work. And of course I'd design the addition.

Int: Could you add a motor generator so we'd have an
alternate power source?

Luis: Sure. We have one now in one of the lab buildings; our people could install one.

Next you decide to discuss your problems with Claire Brunner,
the Institute's safety officer. You're particularly interested in the
fire extinguisher problem.

Interviewer: Claire, I'm really concerned about fire
safety in the data processing center. We've only
got hand-held extinguishers in there now.

Claire: I know. Those were put in a while ago, before
I was hired. The problem is that the center is in a
building that doesn't have a sprinkler system so
those extinguishers are the best that we can do
under the circumstances.

Int: Suppose we were to build a new computer center,
either as an addition to the building or as a separate facility; what kind of system would you suggest?

Cl: Ideally? A halon extinguisher system would be
best. You don't want a water system around electronic equipment and CO_2 is too dangerous to personnel. A halon system for a full-size computer
room would probably run us around $15,000, but it
would be worth it if it saved us from having the
whole computer facility wiped out in a fire.

```
Int:  Could you put that system into a building addi-
      tion or would it require a whole new building?
Cl:   Oh, you could build it into an addition.  It´s
      self-contained.
```

Finally, you consider the location of the computer rooms as they currently exist. The room housing the Digicom is at the north end of the first floor of the administration building; it contains all the equipment associated with the Digicom as well as two desks for the computer operators. Your office is located in a cubicle on the right side of the room. The RDS is in the room to the left; it contains all of the RDS equipment along with a desk for the computer users.

On the same side of the hall as the computer rooms there are a conference room and three offices housing accounting personnel. On the other side of the hall is the building's lunch room. Next to it are four more offices housing part of the payroll department.

Assignment 1. Write a proposal prospectus to the Institute's director of operations based on the information you've collected. Remember, this is not a full-fledged proposal, just an opportunity to get the reactions of others to your ideas.

Assignment 2. The director was impressed by your discussion. Using the information you've collected in your various interviews, write the proposal. Remember to consider appropriate graphics. Your audience is the Institute's Board of Directors. Your purpose is to convince them to build an addition for a new, expanded computer facility.

Assignment 3. The Chairman has decided that you should present your proposal to the board in person. Write the script (audio and visuals) for a fifteen-minute oral presentation of your proposal for the new computer room.

MERIT HOSPITAL: ALCOHOL AND SUBSTANCE ABUSE

You are an assistant hospital administrator at Merit Hospital. Your supervisor, the chief administrator, has asked you to prepare a short proposal—a prospectus, really—dealing with a problem that has surfaced at Merit Hospital. A nurse with fifteen years of seniority was found asleep in one of the patients' bathrooms during her shift. From her condition, it was clear that she had been drinking heavily before reporting to work. In an abstract way, you know that the health care profession is one of the professions with the highest levels of alcohol and substance abuse (Plant 1979). In the last two years, there have been six similar incidents at Merit Hospital, but the others did not raise such a storm because the nurses slept out their shifts where patients did not find them. In a conversation with the nurse in question—who now is on probation —you realize that her case is typical. She has been supported for years by a pattern of collegial support, that is, collegial refusal to recognize the problem or its dangers. You also know that your administrator is not at all keen to tackle this underlying problem of looking the other way and pretending that nothing is wrong.

Last year an emergency room physician at a neighboring hospital died in the staff shower from a combination of the hot shower and the dose of amphetamines and illegal drugs he had just taken to get through his shift. After the initial flurry of news coverage, interest waned and no steps were taken in the medical community to deal with the issue. Yet you realize that people who abuse alcohol also tend to abuse other substances, and you know that drug inventory and drug availability are perennial problems in hospital administration. No one looks forward to federal reviews of drug stocks. No one wants to face the monetary or human costs in terms of lost time, sick time, illnesses, and danger to patients.

Your first step, after interviewing the nurse on probation, is to instruct your student intern—a master's candidate in hospital administration—to put together a review of the literature related to alcoholism among nurses. Since you are not sure yet what proposal you want to make, you don't ask for the literature to be focused in any way. What follows is an abstract of what you receive from your intern.

Alcoholics function at 67% of capacity according to Trice and Roman (1979). In a 1985 study (Iglauer) of 275 nurses who filled out self-evaluation questionnaires, it was found that 12% abstained from alcohol entirely, 42% were light drinkers, 31% were moderate drinkers, and 15% were problem drinkers. Iglauer compared her results to the 1981 Wilsnock study of the general U.S. population, which used the same criteria for categories of drinkers. Wilsnock et al. found 29% abstained, 42% were light drinkers, 22% were moderate drinkers, and 7% were problem drinkers.

Drinking behavior is reinforced and defined by the work climate. There is pressure to conform and help in covering up. Nurses and doctors commonly feel that they can diagnose and treat themselves and resist seeking help about personal problems. When doctors diagnose alcoholic medical staff (their colleagues), they seldom diagnose alcoholism itself. Instead, they prefer to diagnose symptoms, such as sleep disturbance, indigestion, depression, hepatitis, or pancreatitis. Treatment is also generally symptomatic rather than addressed to the root cause. Drinking during pregnancy can cause spontaneous abortion, low birth weight in the child, and fetal alcohol syndrome with effects that are similar to Downs' syndrome. Fifty percent of Americans who drown or die in falls have been drinking. According to other studies, female registered nurses have twice the drinking problem of generally employed women

in the United States. The nurses´ drinking problem is
the same for all types of nurses and all nursing
assignments; nurses on night shift, in managerial positions, or in critical care do not appear to have a
higher incidence of alcoholism than other nurses
(Iglauer 1985).

Zuhorek (1981) documented poorer work ability among
alcoholic medical staff. Jaffe (1982) shows a strong
commitment of the medical profession to initiating
alcohol treatment and awareness programs since the
early 1970s. The U.S. Department of Health and Human
Services Fourth Special Report of 1981 demonstrated
that alcoholism is treatable.

In hospitals, where drugs are used to cure illness,
there is a general climate that supports the use of
alcohol as a "coping drug." Nursing education does not
address the potential for alcoholism or other drug
abuse on the job. Trice and Roman (1982) found that
few of the established substance abuse programs in the
medical profession are aimed at female-dominated areas
of health professions. Bissell (1981) estimated that
there are 75,000 alcoholic registered nurses, extrapo-
lating from the government estimate that 5% of all
American women are alcoholics. Thirty-five to sixty
percent of all auto accident fatalities involve alcohol. A problem drinker is defined as someone who takes
more than 13 drinks each week. By this standard, 4% to
7% of American women are problem drinkers, while 12% of
American men are problem drinkers. Eighty percent of
suicides have been drinking immediately before killing
themselves.

Compared to the total population, a higher percentage of health care professionals were raised in homes
with one or more alcoholic parents (Finley 1982).
Registered nurses were often burdened by responsibil-
ities as children, and this experience influences their

173

decision to choose careers as care givers. Less than
1% of medical school time or curriculum is directed to
issues of alcohol or substance abuse (Pokorny and Solo-
mon 1983).

After reviewing all of this material, you decide that Merit Hospital should offer an institute or workshop on alcoholism awareness. You estimate that with a staff of 500, it will take 10 workshops (50 staff each). Each workshop will take 2 days. Staff on duty can cover for those who are at the workshops, so additional labor costs can probably be avoided. An outside consulting firm offers workshops for $1,000 per workshop. This price includes lecturers, trainers, and materials. You need to design some kind of follow-up evaluation, including a study of patient complaints, absenteeism, absentee excuses, and perhaps a formal survey that might be administered by another masters' candidate. In addition, you think that it will be necessary to send at least 7 percent of your staff to substance abuse clinics. The medical insurance you provide for your staff will cover the cost of the clinics, but it will be necessary to hire replacement staff because each clinic session requires 2 weeks in residence. That's 80 hours off per person, with replacements costing $15 per hour.

Assignment 1. Write a two-page proposal abstract for discussion at the next hospital management meeting.

Assignment 2. Keeping in mind that your audience will be hard to persuade, write the short proposal for the workshops, medical care program, and evaluation measures. Limit your presentation to prospectus or short proposal length—no more than five pages.

Assignment 3. Carry out research similar to the intern's for another field of work and write a similar proposal. You might study alcohol abuse among engineers or chemists, for example. Or, with the approval of your instructor, you might focus more generally on substance abuse (including the abuse of prescription and non-prescription medications and illegal drugs).

PRIVATE PLACES AND SPECIAL PLACES

As a senior majoring in landscape architecture you have become more aware of your surroundings than ever before. As you look around your campus you are struck frequently by problems you never noticed. For example, it occurs to you that there are lots of open areas around the campus buildings but few places where students can sit or work outside, even in good weather. As you consider the problem, you notice that some of the public areas are congested with people and cars while others lack seats or shade or drainage. Some areas are simply too public to be inviting.

You decide that your senior project will be to convince the college administration to develop a particular outdoor spot on campus for study, conversation, or even class meetings on pleasant days. After some consideration you arrive at some criteria you know will interest the administration:

- *Initial cost:* You know the administration won't undertake a project that is too costly. You decide your object will be to redesign or add to a present area rather than digging up an area or building an entirely new facility.
- *Security:* You want to create a sense of privacy without constructing a large fence or totally obstructing the view.
- *Maintenance cost:* You assume that the administration will expect strict controls on the costs of maintaining any campus additions.

Assignment 1. Investigate your campus for an appropriate area that could be developed in the way described. Write a proposal abstract or a short persuasive letter to be published in the school newspaper explaining what you want to do and why it is important. If you wish, you can include costs and consider possible objections. Your purpose is to generate feedback from interested persons, in this case other students. Remember, this abstract or letter is not a finished proposal.

Assignment 2. Student reaction to your abstract is very positive. Obviously you have discovered a real need. Write the proposal itself, directing it to the administration and remembering the criteria you developed earlier.

THE PLATING TANK

For the past four years you have worked as a chemical engineer for Lovett Aircraft Company in its maintenance and quality control division. Your section, maintenance, is concerned with maintaining the several thousand aircraft that fly in and out of Lovett every month. One of the routine maintenance procedures is plating various parts of the aircraft engines with metals, frequently cadmium, to reduce wear and corrosion. This cadmium plating takes place in large tanks containing cadmium oxide, sodium hydroxide, sodium cyanide, and sodium carbonate. Ideally there should be from one to eight ounces of sodium carbonate for every gallon of plating solution. Unfortunately, as a result of the plating process and exposure to air, the concentration of sodium carbonate continually increases in the plating tank. If the concentration goes above eight ounces per gallon, the plating will be affected; the plating may not stick to the metal or it may be pitted. For this reason the concentration of sodium carbonate in the plating tanks must be periodically lowered. At the moment, the method for lowering the concentration involves adding barium cyanide to the tanks; the barium cyanide reacts with the sodium carbonate to form a solid precipitate: barium carbonate. This precipitate can be removed manually.

This method does lower the amount of sodium carbonate, but it creates other problems. First of all, a large amount of barium cyanide is required: almost 500 pounds. The size of the amount slows the reaction itself; it takes around three days for the reaction to be completed. After the reaction has taken place, the plating solution is filtered to remove the precipitate; the filtering requires another two to three days. When the solution has been removed, a work crew must shovel the precipitate out of the tank so that it can be sealed in 55-gallon drums and disposed of. The precipitate is toxic, and members of the work crew must wear protective clothing while removing it from the tank.

You've been trying to come up with an alternative method that would be faster and less dangerous for getting rid of the precipitate, and you think you may have found one. Your method depends on a difference in the amount of sodium carbonate that remains dissolved in the plating solution at lowered temperatures; in other words, a difference in solubility. At 72°F (22°C, roughly room temperature), about 25 grams of sodium carbonate can remain dissolved in 100 cubic centimeters of solution, but when the temperature drops to 32°F (0°C), only 7 grams remain dissolved in 100 cubic centimeters. According to your figures the solubility goes like this: 50 g/100 cc at 40°C; 40 g/100 cc at 30°C; 25 g/100 cc at 22°C (room temperature), and 7 g/100 cc at 0°C. The solubility drops along with the temperature, and the sodium carbonate forms on the coldest surface of the solution. In this method, the temperature rather than the barium cyanide causes the precipitate to form.

Now you need to come up with some way to use the lowered temperature method on the plating solution. After doing some research, you discover a design that might work. First, the workers would pump the plating solution into a cooling tank containing a cooling coil pumped full of antifreeze. Such cooling tanks already exist in the plating shop where they are used for other purposes; thus no new equipment would be required. Next the temperature in the tank would be lowered to $32°$ to $50°F$ ($0°$ to $10°C$), and the solution would be left in the tank overnight. By your calculations, enough sodium carbonate will form on the cooling coil during the night to bring the concentration in the plating solution back to the right level. The plating solution would still have to be filtered, but the filtering would take only one day. Then the solution could be returned to the original tank. The antifreeze would be removed from the cooling coil and the cooling tank would be filled with tap water, which would be around $68°F$ ($20°C$). The tap water would cause the sodium carbonate left in the tank to dissolve again. Then the solution of water and sodium carbonate could be pumped into a tank truck and removed.

Now that you have your method, you need to investigate costs; this is what you come up with.

```
Costs:
Method 1
Barium cyanide, 500 lb @ $10/lb ................$5,000
Two maintenance workers, 8 hr @ $9/hr/worker.....   144
One 55-gallon drum...............................   103
Method 2
Two maintenance workers, 3 hr @ $9/hr/worker.....    54
Tank truck use...................................    50
```

Assignment 1. Write a proposal abstract for the new method of removing sodium carbonate from the tanks. Your audience is your section chief, K. F. Singh. You want to explain the method and get his reaction before you submit the proposal formally.

Assignment 2. Mr. Singh likes the idea very much. He suggests that you submit the new method as a proposal to the chief of the maintenance division, Roy Kubitski. Write the proposal. Your purpose is to convince the division chief that the method should be implemented. Consider what points would be of most interest to him; also consider what graphics you would include. Here are some technical terms you might want to use.

```
Symbols:
Barium carbonate:  BaCO3
Barium cyanide:  Ba(CN)2
Cadmium oxide:  CdO
Sodium carbonate:  Na2CO3
Sodium cyanide:  NaCN
Sodium hydroxide:  NaOH

Solution reaction:  Ba(CN)2 + Na2CO3 = BaCO3 + 2NaCN
```

THE WRITING CONFERENCE

For six months you have been a junior manager in the executive office of Fairfield Office Equipment, a large manufacturer and retailer of office furniture, business machines, work stations, and office supplies. Fairfield's staff is characterized by a great diversity of professional training, level of education, and understanding of the business. Moreover, the company's facilities span eight states, including work station and computer divisions in Boston and Chicago, steel fabrication factories in Ohio and Indiana, furniture design and manufacturing centers in the Carolinas, and several warehouse and sales operations in the west.

Fairfield managers have always worked to maintain clear communications through centralized training, quarterly regional meetings, and an annual meeting of upper management that is held at a different Fairfield factory each year. In a cycle of eight years, a manager will personally visit each major facility. Recently, however, corporate communications have been breaking down.

Last week, Margaret Fuller, your immediate superior and the Assistant Vice President for Product Development, called you in to show you the latest evidence of miscommunication. The furniture division in North Carolina built a trial run of a computer work station desk that does not provide sufficient space for ventilation, has no room for the keyboard cable, and has only a tiny hole at the back for a power cord. Consequently the unit requires extensive modifications to allow easy connection of cables for a video screen, modem, network, printer, or other auxiliary systems. Furthermore, no provision was made in the work station for printers that feed paper from the bottom.

After reviewing the various memos and drawings traded by the two divisions responsible for the design, Ms. Fuller is convinced that the problem is more than lack of clarity. Each division made unwarranted assumptions about the other division's understanding and knowledge; key memos were obviously written both in haste and with reluctance; communications were hard to follow—often sloppy and badly organized.

Ms. Fuller asked you to put together a writing conference for the Chicago headquarters; if the conference is a success, then a similar one will be given at other large plants and sales centers. Before she sent you off with your assignment, she brainstormed with you for half an hour while you took rapid notes on each topic as it came up. Here are your notes.

```
This has to be more than just another conference.
Remedial?  Invite only people who can't write?
Best people/Best writers to serve as models?
Invitational or compelled attendance?
Remedial conference would send out a negative signal
about writing.
```

```
If the objective is to build on the good communi-
cators, then the company needs to follow up.  Good com-
pany newsletters; discussion of writing in company
meetings; model communications from the top down.

What is this going to cost us for a single pilot work-
shop?   $500 for a one day consultant--say 6 hours.
That´s about all anyone can handle in a day.
```

At this point, Ms. Fuller sent you off to do your own research and to prepare a proposal for a one-day workshop for about twenty people. Your proposal is to include a budget. During the next week, you call a consulting firm and the local university to find out who runs corporate writing meetings. You come up with the following additional information.

```
A writing workshop can help to promote both communica-
tion and productivity.

Your conference could cover some or all of the follow-
ing topics:   oral communication, computerized word
processing, editing other writers´ work, electronic
mail, clarity, basic grammar, quick outlines and memo
formats, good organization, sharing sessions to talk
about divisional problems.

Ed Zoster of Corporate Communications Experts will do
workshops on writing with computers for $75/hour.  He
is strongly enthusiastic about electronic mail systems
that connect computer work stations and provide staff
with instant mail.  He can also talk about the use of
quick outlines to organize memos and short reports
written and mailed on a computer--without any revision.
Ed also advocates computers for slow writers because
some research suggests that computers are less intimi-
dating than empty sheets of paper for some blocked
writers.
```

Mary Wade, from the local university's writing center, suggests a more person-centered approach to the workshop. She emphasizes the importance of everyone sharing concerns and writing out their objectives at the beginning of a workshop. Her consultants can provide short segments on organizing, outlining, audience analysis, and standard memo-report formats. She also believes that workshops should be reinforcing. Each attendee should go back and teach one other person, and every attendee should feel valuable and even privileged to be part of the workshop. For that reason, Mary says that everyone should be treated to a good lunch, preferably at a site away from the company, and a small sheaf of handouts, such as suggested outlines. She also cautions against expecting a quick fix from a single conference.

Mulling all these data, you call around to get some facts about communication patterns in the Fairfield regional offices and factories. In Chicago and Boston, about 45 percent of all managers are connected to computer networks. Corporation wide, about 10 percent of the employees have access to networks. About 22 percent of all offices make extensive use of word processing systems for correspondence and reports; the rest are still using electric typewriters or electronic typewriters that have limited memory and correction systems.

The budget office provides you with additional data. Lunch costs about $9.00 per person at the local Greek restaurant. Use of the corporate bus costs $1.00 per person. Two-color quality handouts on glossy paper cost about $1.00 each for a press run of 500 plus a design cost of $35. Photocopying the same materials in-house will run about $10.00. You probably need two different handouts. Morning and afternoon coffee and snacks, delivered by the company concessions office, will cost $2.25 per person. Finally, you estimate that each person who comes to the workshop will represent a division or work group of 10 to 125 people for an average of 24.

Today you returned to Ms. Fuller's office with a preliminary oral report on what you have done. After hearing your report, she asks you to prepare three documents: a written proposal to her—including a schedule for the conference, names of presenters, and a budget—and two letters. The first letter will explain the conference to division managers; the second letter will be an invitation to be sent to the first group of twenty participants. As an afterthought, she asks you to be certain to include some form of evaluation at the end of the conference.

Assignment 1. Oral presentation: After scanning the preceding data and perhaps doing some additional research or using data provided by your instructor, you realize that the conference will probably cost more than $500. Design a workshop and prepare a budget. Then make the oral presentation to Ms. Fuller, explaining what you have done and your conclusions. Make this presentation to your class.

Assignment 2. Write the proposal to Ms. Fuller, persuading her to accept your workshop design. Include a full workshop schedule and budget.

Assignment 3. Write the letter to the divisional heads. Remember that you are persuading them of the importance of the workshop as well as informing them about it.

Assignment 4. Write the letter of invitation. Ms. Fuller has decided that the workshop should focus on building the skills of strong communicators so the participants can go back and teach what they learn to others in their divisions. Since many of these people are highly valued in their divisions and busy with significant projects, you must make the workshop appear to be a good use of their time.

Chapter *7*

BROCHURES

Brochures call for careful integration of text and visuals, careful audience analysis, and particular attention to the circumstances in which they are likely to be read.

CHARACTERISTICS OF BROCHURES

Brochures' text should consist of short sentences and short paragraphs. Since brochures are often skimmed quickly with the mail or read while walking around a store, they should be designed for skimming and intermittent reading. Those conditions call for frequent subtitles or headings—and short sections. If a reader needs to look up for a second, it should be easy for her to find her place again.

Brochures rely heavily on visual impact. The pictures and text should be combined so that the reader is guided *to read the text* and *from one section of text to the next*. If you study professionally designed brochures, you'll probably notice that special efforts have been made to assure that when the cover of the brochure is opened, the reader will start reading the first panel of information, not some other section. Visuals are often designed so that the reader's eye is directed in a pattern that starts with the top of the text, moves toward the last panel of text, and then sends the reader back to the beginning for another look. Often this circling effect is achieved by using a photograph or drawing that clearly includes a circular design that points back toward the first panel.

Lay out a brochure so that the cover directs the reader inside and the inside conveys the message. Use headings and visuals to ensure that the reader sees all the text. Use the back for information of minimal importance and for mailing labels. Figure 7-1 illustrates several features of the design of a simple six-panel, 8½" x 11" brochure.

Brochures call for particularly careful analysis of audience and purpose. When you have only a few words to use, you can't afford to lose an audience because the language is inappropriately technical. You must also be aware of all the possible misinterpretations of metaphors and visual images. Many readers may stop after a single panel, and every word and panel must contribute as much as possible toward persuading and informing. Brochures are often skimmed, not read.

STEP-BY-STEP BROCHURE DESIGN

After assessing the likely audiences and purposes, you should consider how the brochure will be distributed. Will it be a hand-

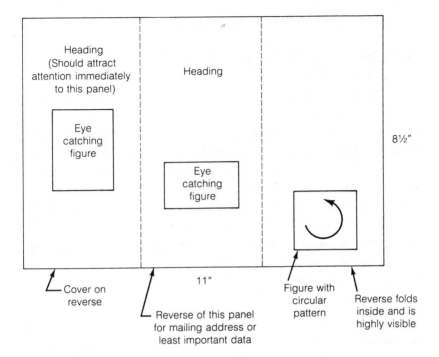

Figure 7-1. Sample brochure design

out? Mailed? Packaged with other materials, such as an educational or promotional packet passed out during a convention? Where will it be distributed? The answers to these questions should shape your design decisions.

Usually a brochure aims to create a single major impression and to prompt a single action. Brochures are not a good medium for conveying multiple purposes or reaching multiple audiences.

Next consider what formats are possible—given your decisions about audience, purpose, and budget for production and distribution. Finally, you can write the copy, keeping in mind your readers' need for short and simple copy that can be read anywhere.

FOLDING THE BROCHURE

In addition to the basic two-fold brochure illustrated in Figure 7-1, you can try several other common styles—keeping in mind that whatever you produce should fit into standard sized envelopes if you are designing it to be mailed. You can fold a standard sheet of paper once, creating four panels, and let the pamphlet open sideways like a book or up like the lid of a chest. The accordion fold is another alternative. Before you begin work on any of the assignments in this chapter, collect five or six brochures from your school, or from banks, government agencies, and markets. Studying examples will teach you much about the details of layout and the uses of color, photography, headings, and other visuals, as well as about the language and style appropriate for this compact medium.

CASES

THE AQUIFER

For the past five years you have worked as a geologist for the Osage Underground Water District, helping to analyze and map your region's chief source of water: the Osage Aquifer. The aquifer is a 175-mile long underground reservoir extending under four counties and parts of two more. Although many people refer to the aquifer as an underground lake, it is actually a massive formation of water-bearing rock—a huge limestone deposit.

The aquifer has served as a water source for your region for thousands of years, and, until the last decade, local authorities considered it sufficient for years to come. But over the last ten years your area has experienced an economic boom. Widespread expansion has brought new industry and several thousand new residents; the demands on the aquifer have risen sharply.

Most people were not greatly concerned about the aquifer until this year, however. During this summer your region experienced its worst drought in seventeen years. The water level in the aquifer dropped precipitously, and various city governments have instituted voluntary (sometimes mandatory) water rationing. Suddenly everyone is concerned about the aquifer.

Your boss, the head geologist of the district, thinks that this is an ideal time to educate the public about its primary source of water. Although other sources may eventually be necessary, the aquifer can continue to supply the region's water needs adequately if the population learns to practice conservation. Your boss feels that the best way to convince people of the value of this conservation is to explain what the aquifer is and how it works. He is particularly interested in explaining the aquifer to school children since he feels they can spread the information to their parents and they will remember the information themselves when they grow older. You have a feeling that his decision means more work for you, and you're right.

Your boss's idea is for you to write a pamphlet aimed at an audience of twelve- to fourteen-year-olds, explaining what the aquifer is and what the current problems are. You start by reviewing the information you have on hand.

```
Aquifer Statistics

175 miles long, from Lobel County west to Sedgewick.
Also stretches over McDade, Santa Rosa, Benteen,
Carroll, and parts of Lincoln and Tomal Counties.


Width varies from 5 to 30 miles.


Makeup:  Limestone formations with layers of im-
permeable clay above and below.
```

Recharge zone: Where water enters aquifer; seeps in through faults in impermeable clay layer from creeks, rivers, lakes. Also rain water runoff enters directly through faults.

Runoff rain water primary source of recharge.

Discharge: Water flowing out of aquifer through wells and springs. Wells provide water to main urban areas. Springs supply tributary water for the San Luisa, Ottumwa, and Saba Rivers.

Water Use

Primary uses: Agriculture (crops, livestock, food processing), municipal (residential and commercial), industrial (manufacturing and power generation), recreational (parks and recreation areas), and natural (maintenance of ecosystem).

Farming: Crops require 24" to 36" of water annually. Average yearly rainfall, 20" to 34". Water pumped from aquifer makes up difference. 38% of water pumped from aquifer goes for irrigation.

Municipal: Cities are second largest drain on aquifer. Water use in McDade County: 180 gallons per person per day (includes household, commercial, and some industrial use). 48% of water used in home. Family of four uses 90 gallons on toilet flushing daily. 70 gallons used on bathing. 30 gallons for laundry. 25 gallons for kitchen. 10 gallons for cooking and drinking.

Current Problems

Current drought situation: No rainfall = no aquifer recharge. Water level in aquifer dropping.

Irrigation demands (agricultural and residential) higher because of freeze last winter (destroyed crops and lawns). People trying to regenerate plants.

Lack of rainfall and high summer temperatures also hard on vegetation.

Record aquifer low set 30 years ago: 612 feet. Currently 624 feet. Rain could bring level back up.

Continued drought and dropping aquifer level could lead to curtailment of agriculture (major occupation of region), curtailment of recreation (towns along Saba and San Luis Rivers depend on summer tourism dollars), and curtailment of growth.

Future

Population growing approximately 40,000 per year; expected to reach 1.5 million by year 2000.

Aquifer discharge exceeded recharge past three years.

Conservation needed both on large scale (industrial and agricultural) and small (residential).

Assignment 1. Write the pamphlet on the aquifer, using the information and organization you think would appeal most to an audience of twelve- to fourteen-year-olds. (The pamphlet will be distributed in middle schools and junior high schools.) A secondary audience is the children's parents. Remember to consider both style and graphics appropriate to your purpose.

Assignment 2. Your boss likes the original pamphlet so much that he wants you to write another on the same subject. This time your audience is the general public. The pamphlet will be distributed through participating supermarkets and banks. Your purpose is to inform the population of your region about its primary water source.

SMOKING AND THE PILL

You have been hired by the board of a local birth control clinic to prepare a brochure that explains to the appropriate audience of patients the dangers of taking birth control pills while continuing to smoke. Your task is to decide what part of the clinic's clientele you wish to target and then design a brochure that communicates the information.

You must look the data up in the *Physicians' Desk Reference,* the standard trade guide to medications. The *PDR* is available in most university libraries, health center libraries, and medical offices. Look up contraceptive pills or use the brand name index to find the page numbers for information on common brands of contraceptive pills, such as OVCON. You should find a four- to six-page discussion with several charts and graphs. Use the information you find there to write the text of the brochure, adapting it to your audience, the clinic's patients. Be sure to cite your source and credit your sources for any graphs or data you reproduce.

Once you design the brochure, your job is not done. You must also write a letter of transmittal to the board of the clinic, explaining why you designed the brochure the way you did. You should be particularly clear about your choices of audience and purpose, and you should be as objective as possible. The board of the clinic feels that the information in the *PDR* should be made available to patients, but they want to leave the decision to smoke or not to each patient.

Your primary task is to make the data in the *PDR* accessible to patients. The *PDR* is written for an audience of physicians, nurses, and other medical professionals.

Assignment 1. Write and design the brochure.

Assignment 2. Write the explanatory letter of transmittal.

CHOOSING A CALCULATOR

As president of your university's engineering club you've been trying to organize programs for new engineering students. Because of the size of your engineering department, many of the new majors feel somewhat lost and confused about the demands their career will place upon them. You've already put together a pamphlet about the courses that are recommended for the individual engineering specialties, but now you'd like to do something to help with other choices. You're trying to put together pamphlets on the various types of equipment the students may need to purchase during their academic careers; you'd like to discuss reference manuals, drafting sets, calculators, and even home computers. You're going to do the pamphlet on calculators yourself since you just purchased one after an extensive search.

As you get ready to write the calculator pamphlet, you go over the notes you made when you were trying to decide which one to buy.

```
Notes
Possibilities: Wide variety available--financial,
checkbook types, scientific, programmable, even hand-
held computers. Probably scientific types best for my
purposes; others have things I probably wouldn't use
(e.g., all that percentage stuff on the financial).
```

<u>Scientific</u>: Nonprogrammable, programmable, expandable programmable. Expandable can have printers and external program storage.

<u>Hand-held computers</u>: No. According to the salesman (J. Friedman) at Quality Electronics, if a computer is needed, better use a regular one.

<u>Features</u>: Display (i.e., number of digits it can work with). More digits the better. Mantissa (in scientific or engineering notation, part of number described by exponent)--ditto, more the better. Memory--how many numbers can it store? More than one memory helpful since you may have a lot of constants to keep track of. Also should have constant memory that remembers numbers when you turn it off. Statistics capabilities are nice for statistics equations; not strictly necessary though.

<u>Functions</u>: Functions = operations done on a key. More functions, fewer keystrokes, more time saved. Also less chance of mistakes--less chance of hitting wrong key if function is automatic. Functions needed: Basics (i.e., =, -, x, etc.), trig functions, log functions, y^x, $1/x$, change sign, exponential, x exchange y, summation, and degree/radian/grad. Other possibilities: rectangular-polar conv, standard deviation, hyperbolic functions, engineer notation, random number, degree (to hr, min, sec), fraction, round number, percent change, insert/delete, clear.

<u>Costs</u>: How to compare? Some have more and cost more. Ideal: Most for the money. Maybe cost per function (i.e., divide price by number of functions). Sample costs in this area: nonprogrammable, $30-$40. Cost

per function, .50-.75. Nonexpandable programmable,
$40-$70. Cost per function, .35-1.00. Expandable
programmable, $65-$150. Cost per function, .15-.20.
Programmable: cost divided by available program lines.

Assignment. Write the pamphlet giving advice on choosing the best calculator. Your audience is made up of first year engineering students who may or may not know much about available calculators.

ANTIBIOTICS IN ANIMAL FEED

As an intern with the local consumer cooperative, you have been asked to research an issue that has been brought before the board of directors. Lately, some members of the cooperative, which operates a grocery store, have become concerned about the use of subtherapeutic doses of antibiotics in poultry and livestock raised for food. A subtherapeutic dose of an antibiotic is less than the amount that would be prescribed to fight an infection but high enough to depress the levels of bacteria in the animal. Antibiotics include drugs like penicillin and tetracycline. Some experts argue that because large quantities of antibiotics are being used to prevent disease among animals that are kept in cramped feeding pens, the bacteria of the world are gradually becoming immune to the action of all the major antibiotics. The co-op board asked you to put together an informative brochure about the use of antibiotics in animal husbandry. You are not to take a strong position on either side of the issue. At this stage, the board prefers that you introduce some of the reasons that antibiotics are used in animals and poultry raised for food, describe the extent of their use on a typical farm, and provide an overview of the general issue.

The following facts include all the data you discovered after several interviews and some library research.

1. Interview with nutritionist on duty in the store. Americans should eat only chicken because the antivivisectionist movement abhors the way that other animals are raised and slaughtered.

2. Remainder of data from phone call to state agricultural extension agent and visit to library. The Food and Drug Administration allows animals to consume chlortetracycline up to the day they are processed. Administration of other drugs must be stopped anywhere from a few days to several weeks before slaughter. F.D.A. guidelines vary from drug to drug. (Who sets guidelines?)

3. A typical therapeutic dose of tetracycline for an adult human is 1 g to 2 g per day over a period of 5 to 7 days. (Is this adjusted to weight? Condition? Disease?)

4. Pig breeding typical situation for animal use of antibiotics. Newborns (piglets) injected for survival in farrowing crate. Spend first 4 weeks in close quarters in farrowing crate with sow and litter mates. Weight increases sixfold in this period.

5. From 40 to 70 lb pigs eat 1/5 lb antibiotics per ton of feed. After this, pigs eat 1/10 lb per ton of feed.

6. Large piles of manure (animal waste) produce large volumes of ammonia. Ammonia damages lungs--increases chance of pneumonia, other lung diseases. Low doses of antibiotics help cut down lung disease epidemic. Dangerous bacteria etc. spread quickly in cramped barns, feeding stalls, can cause dysentery, tissue and bone disorders. Low doses of antibiotics keep these under control.

7. In 6-month growth cycle, pig may eat 20 g of medications.

8. Antibiotics given in subtherapeutic doses. Dosage not enough to cure pneumonia or other diseases. Enough to help reduce outbreaks of disease in a population. Cramped barns ideal conditions for rapid spread of disease.

9. Sow during nursing given 1 part antibiotic to 2,000 parts food (1 lb per ton?). This is 1/10 of therapeutic dose. From the time a pig is 12 lb to 40 lb, daily ration contains 1/10 lb medication per ton of feed.

10. Some farmers treat diseases by removing animals to an open field. Space, clean dirt, sun and air.

11. Studies in past decades showed antibiotics in low doses increase weight gain of livestock and poultry. Recent studies question earlier results. Animals kept in pens can be monitored more closely than range stock. (Which studies right?)

12. Good farm land costs several thousand dollars per acre. Too expensive to range livestock?

13. Cost of meat if intensive farming practices changed?

14. Most common strains of bacteria (E. coli, for example) now unaffected by antibiotics. Resistance to antibiotics can be quickly transferred between species of organisms. Antibiotic regimens (subtherapeutic doses) banned in some nations. Salmonella outbreaks.

15. What do we eat? What happens to antibiotics in processed meats? (Are they in the meat? Can resistance be transferred by eating meat?)

Assignment 1. Write the brochure. Note: Your instructor may want you to do more research on this question. The final brochure is to be designed for an audience of intelligent and concerned consumers; it will be made available at a public information display inside the grocery store. It should be limited to a single folded sheet of 8½" x 11" paper and include at least two illustrations.

Assignment 2. The co-op board wants more information. Prepare a short research report on the most recent information on this controversy. Your sources should be less than one year old.

Index